DIFFUSION-DRIVEN
WAVELET DESIGN
for SHAPE ANALYSIS

DIFFUSION-DRIVEN
WAVELET DESIGN
for SHAPE ANALYSIS

TINGBO **HOU** • HONG **QIN**

CRC Press
Taylor & Francis Group
Boca Raton London New York

CRC Press is an imprint of the
Taylor & Francis Group, an **informa** business

CRC Press
Taylor & Francis Group
6000 Broken Sound Parkway NW, Suite 300
Boca Raton, FL 33487-2742

© 2015 by Taylor & Francis Group, LLC
CRC Press is an imprint of Taylor & Francis Group, an Informa business

No claim to original U.S. Government works

Printed on acid-free paper
Version Date: 20140908

International Standard Book Number-13: 978-1-4822-2029-2 (Hardback)

Visit the Taylor & Francis Web site at
http://www.taylorandfrancis.com

and the CRC Press Web site at
http://www.crcpress.com

To my family.

Contents

Preface

Wavelets are powerful tools in functional analysis and many engineering fields. In the community of computer graphics, we are thrilled for the success of wavelets on three dimensional shapes, like what they have done in image processing.

About the Book

This book collects fresh research results in wavelet designs on three dimensional shapes. The proposed methods can be applied to various types of three dimensional data: polygonal meshes, point clouds, manifolds, and volumetric images. The design is driven by heat diffusion. The heat diffusion process is a natural phenomenon that connects local and global geometry in a multiscale way. It builds a multiresolution representation of any value function defined on the domain. It is computable on manifolds with nice formulation. Therefore, it greatly facilitates the design of wavelets on three dimensional data.

Through this book, we are eager to introduce the design methods of wavelets on manifold data, and advocate the applications in shape analysis. Designing wavelets on manifold data connects knowledge in differential geometry, functional analysis, Fourier transform, spectral graph theory, and stochastic process. For a given shape, the wavelets $\psi_t(x, y)$ are functions with points x, y in the spatial domain and time t in the frequency domain. They are purely determined by the shape geometry. Wavelet transforms are computed as inner products of wavelet kernels and input functions. The second part of the book presents various applications. As intrinsic functions, wavelets characterize shape geometry, and therefore can be used for shape representation. As spectral tools, wavelets are used for geometry processing with filters in a joint space-frequency domain. As detail extractors, they are used for shape feature definition and detection. Beyond these fundamental applications, we also introduce some middle-level and high-level applications, including shape matching, shape registration, and shape retrieval.

Unlike many other wavelet books, this book does not involve complicated mathematics. Instead, it uses simplified formulations and illustrative examples to explain deep theories. The book is based on very recent research results in computer graphics, so readers can reach to the frontier of this field.

This book is valuable to researchers in computer graphics related areas, by means of collecting recent research results in wavelet design and application. These results made significant contributions that facilitate space-frequency analysis of wavelet transforms on manifold geometry. Besides, some unexplored topics are prompted to inspire new research directions. The methods and application in the book are valuable for research in computer vision, visualization, medical imaging, and geometric modeling as well.

This book is also valuable to graduate students and senior undergraduate students majoring in computer science. It explains profound theories through easily understood words and simple examples. Implementation details and application algorithms are well documented in this book. We also provide sample codes on the website of this book.

About the Cover

The cover contains two groups of models. The first Armadillo model comes from Fig. 4.6. It demonstrates a wavelet decomposition of the shape. Details are extracted from the shape at different scales, rendered by coded color. The figure includes two applications of this wavelet transform: local geometry processing and feature detection.

The second Lion model comes from Fig. 5.3. It shows some Mexican hat wavelets at a time computed directly on the shape. The coded color represents function values, from which we can see the wavelets are oscillating and attenuated over the domain. At different locations, the wavelets have different distributions due to the geometry. This shows an impression that the wavelets characterize shape geometry. For more details, please refer to Chapter 5.

Online Resources

To help readers understand the contents and implement the algorithms, we provide online resources available at

https://sites.google.com/site/houtingbo/book

The website contains supplementary figures, videos, and sample codes. All the resources are available for public use.

Acknowledgments

The first author wants to thank the collaborators who made great contribution to the research papers that the book is partly based on: Shengfa Wang, Shuai Li, Ming Zhong, and Xiaohua Hou.

Thanks to the Computer Science Department and the Geometric and Graphical Modeling Lab at Stony Brook University, for the academic environment that incubates the ideas and research results.

Writing a book in spare time is never an easy task. The first author also wants to thank his wife who sacrificed her vacation plans.

The data used in this book is collected in part from some research work [Zhang et al. 04, Vlasic et al. 08, Wang et al. 08, Bronstein et al. 10a], and by courtesy of the AIM@SHAPE repository. The authors are grateful to all data providers.

Tingbo Hou
Hong Qin

Mountain View, California
May, 2014

List of Tables

List of Figures

List of Algorithms

1

Introduction

1.1 Wavelets on 3D Shapes

Spectral analysis tools like Fourier transform and wavelet transform have been widely used in many engineering fields. One such example is image processing, where images are taken as real functions on a two-dimensional (2D) domain. Spectral analysis is to analyze and process functions in frequency domains, or in another saying, at different scales. It brought fundamental applications to image processing, such as feature detection[1], smoothing, denoising, and compression. These low-level applications solidly support the high-level problems of the frontier research: object recognition, image retrieval, and clustering, to name a few.

In this book, our interests are beyond images. We consider a three-dimensional (3D) shape that is a common data type in computer graphics. A shape is a curved surface with or without boundaries. 3D shapes carry more rich information and exhibit more attractive visualization than images. With the rapid development of 3D data acquisition, a massive collection of dynamic shapes emerges and becomes ubiquitous in various real-world applications, resulting in the urgent need calling for fundamental techniques like spectral analysis tools that can be applied to 3D shapes.

A 3D shape $M(x, y, z)$ can be viewed as a function on a 2D domain by projecting the shape to a plane

$$P : [x, y, z] \rightarrow [x', y', h(x', y')]$$

and using the heights $h(x', y')$ as function values, just like an image. It looks like we can directly apply the methods on images to 3D shapes. However, it is usually difficult or impossible to find a projection direction without overlaps for a shape, for example a sphere. To overcome this problem, 2D Fourier transform constructed on small local patches of a large shape was adopted in some research work [Pauly and Gross 01]. This, however, still has problems for deformable shapes. The constructed Fourier

---

[1]For example, the well-known scale-invariant feature transform (SIFT) [Lowe 04] is a wavelet method.

transforms are affiliated to a specific embedding of a shape. Every time when the shape deforms, its Fourier transforms have to change as well. Deformation is unavoidable for non-rigid shapes. Fortunately, a large group of commonly-seen deformation is isometric or near isometric. Isometry is a distance-preserving map between metric spaces. Another close concept is the conformal map, which is angle-preserving. On manifolds, the metric is geodesic. Isometric deformation indicates shapes undergoing deformation have the same geodesics. For example, scans of human performances have large natural deformations that are nearly isometric. This calls for intrinsic quantities that are invariant to isometric deformation.

Alternatively, graphics scientists adopted spectral graph theory for frequency analysis by taking 3D shapes as graphs [Karni and Gotsman 00, Kolluri et al. 04, Ben-Chen and Gotsman 05]. In this theory, eigenfunctions of the graph Laplacian compose an orthogonal and complete basis of the function space on a graph, which are also called "harmonics". They are analogous to the Fourier series, and therefore, can be used for spectral analysis. The set of the graph Laplacian eigenvalues is called the graph's "spectrum", which is connected to frequency in Fourier transform.

Later, this approach has evolved to manifold harmonics. Manifold is a topological space that resembles Euclidean space near each point. More precisely, on a d-dimensional manifold M, each point has a neighborhood that is homeomorphic to the Euclidean space $\mathbb{R}^d$. Roughly speaking, 3D shapes without self-intersection are 2D manifolds embedded in $\mathbb{R}^3$. The eigenfunctions of the Laplace-Beltrami operator are harmonics of the manifold [Lévy 06]. An integrable function defined on a manifold can be projected to the manifold harmonics. In [Valette and Prost 04], this was called the manifold harmonics transform, which is an analogous Fourier transform on manifolds.

Different with the Fourier transform, wavelet transform has both localizations in frequency and space. It can extract function details at given frequencies and locations. The development of wavelets on 3D shapes is more difficult. It starts from the subdivision wavelets [Lounsbery et al. 97]. A subdivision scheme iteratively refines a mesh, resulting in a multiresolution representation of a shape. The process is inverse to the classical filter bank of discrete wavelets. Very recently, diffusion wavelets [Coifman and Maggioni 06] and spectral graph wavelets [Antoine et al. 10, Hammond et al. 11, Maggioni and Mhaskar 08] have been proposed in mathematics. They have wavelet transforms defined as inner products of wavelets and given functions, which equip them with the ability of functional analysis. Diffusion wavelets are constructed by dyadic powers of a diffusion operator, in a bottom-up manner. The spectral graph wavelets are constructed by some wavelet generator designed in graph spectral domain. More details can be found in Chapter 2.

Designing wavelets on manifold data is an interdisciplinary topic, involving theories in differential geometry, functional analysis, stochastic process, Fourier transform, and spectral theory. *Differential geometry* uses differential calculus and integral calculus to study problems in geometry. In this discipline, the Laplace-Beltrami operator, defined as the divergence of the gradient on differentiable manifolds, is the differential operator in the heat equation. *Functional analysis* studies spaces of functions, for example the Hilbert space L^2. It equipped the wavelet design on manifolds with fundamental tools: inner product, norm, and distance. *Stochastic process* is a collection of random variables with indeterministic evolvement. A diffusion process, which can be interpreted as a Brownian motion, is a stochastic process. By this point of view, it is easy to understand the stableness of heat diffusion. *Fourier transform* is a functional analysis method that transforms functions to frequency domain by a complete orthogonal basis. It is a necessary tool to study continuous wavelets on manifolds. Considering the profoundness of involved mathematics, we will use simple terms and practical examples to make it intelligible for graduate and senior undergraduate students in computer science. *Spectral theory* uses eigenvalues and eigenvectors of structure operators to process input functions. It is feasible to more types of data, e.g. graphs, meshes, point clouds, and manifolds.

Wavelet design has a closed connection with heat diffusion. A diffusion process defined on a 3D shape can gradually smooth functions on the shape. It naturally connects local and global geometry, which makes it a powerful tool for multiscale analysis. It builds a multiresolution representation of the function, in a nested chain of subspaces. Wavelets live in the complementary space between two subspaces, and wavelet transform captures function details between two smoothed versions at different scales.

1.2 Book Contents

This book documents some recent research results in wavelet design on 3D shapes and their applications in shape analysis. It has two parts: theories and applications. The first part presents some designed wavelets driven by heat diffusion, and extensions of wavelet generation on volumetric data and manifold data. The second part details some fundamental applications for shape analysis, with a purpose to exhibit how wavelets can help solve many problems in computer graphics.

All the design methods in this book are based on the heat diffusion theory, which naturally brings the rest parts of the design together and provides a unified theory for the wavelet designs. The diffusion process on a shape, governed by a partial differential equation (PDE), captures all the geometric information up to isometry. It bonds local and global geometry in a multiscale manner, which makes it a powerful tool for multiscale analysis. It is resilient to noise and small holes, as the diffusion between two points is determined by its all connected paths. It allows us to design diffusion-driven wavelets on manifold data with rigorous formulations. The heat equation is invariant to isometric deformation, hence the wavelets derived from it. Heat kernel is the fundamental solution to the heat equation. Since it was introduced to computer graphics by [Sun et al. 09], heat kernel has been populated to various problems: feature detection, shape representation, shape matching, shape retrieval, etc. In Chapter 5, we will show that the heat kernel can be used to design the well-known Mexican hat wavelet on manifolds.

The designed wavelets are extended from the aforementioned diffusion wavelets and spectral graph wavelets. The admissible diffusion wavelets are discrete wavelets constructed by dyadic powers of a local operator. The powers of a diffusion-type operator simulate a diffusion process on the data domain. The operator can be viewed as a transition matrix of the Brownian motion, which measures possibilities of where a random walk goes. The bottom-up construction is inspired by the diffusion wavelets [Coifman and Maggioni 06], but with different foci. The diffusion wavelets are concerned more with orthogonality of scaling and wavelet bases, while the designed wavelets focus more on affordable space-frequency analysis on 3D shapes.

The Mexican hat wavelet is a continuous wavelet on manifold data. It is derived from the heat kernel, by taking the negative first derivative with respect to time. This formulation is coincident with its original definition in Euclidean domain, by noticing the heat kernel is a fundamental solution to the heat equation on manifold data. More importantly, this formulation reveals a momentous wavelet design in Fourier domain. It is an independent work with a similar idea of the spectral graph wavelet in [Antoine et al. 10, Hammond et al. 11]. Functions without closed-form expressions may have simply analytic expression in Fourier domain. For example, the Mexican hat wavelet has an analytic expression in Fourier domain: $\lambda_k e^{-\lambda_k t}$.

The design of the anisotropic wavelet is a compound of the above two schemes. We adopt an edge-weighted heat kernel for anisotropic diffusion, based on normal-controlled coordinates [Wang et al. 11a]. Yet, in theory, any anisotropic diffusion kernel or operator would fit in the design. Anisotropic diffusion has better control on directions of heat diffusion. Upon the design, it can suffice different application needs, for example to underline sharp edges of shape geometry. Given an anisotropic diffusion kernel, we can design wavelets by the matrix power scheme or the Fourier transform scheme.

The second part reveals some fundamental applications of designed wavelets for shape analysis. It exhibits how wavelets can help solve problems of 3D shapes in computer graphics. With particular interests, we endeavor to broaden the application scope of wavelets to shape analysis. In this book, shape analysis is referred to a wide range of problems in computer graphics for analyzing, processing, and synthesizing 3D shapes, for example shape representation, feature detection, saliency visualization, shape matching/registration, shape clustering/segmentation, shape retrieval, geometry processing, multiscale approximation, etc. In these applications, we advocate wavelet transforms and their abilities of space-frequency analysis, as how they succeed in other engineering fields. The introduced wavelets are bivariate kernels defined on meshes, point clouds, and manifolds. For a given function, its wavelet transform extracts details of this function at specified location and scale, by the inner product of the function and the wavelet. This is very useful in visualizing saliency information and finding features on the shape. Following this way, a function can be decomposed to a set of wavelet coefficients at different scales. Specifically, we design the decomposition to be complete. Therefore, the function can be recovered from its wavelet coefficients. Moreover, modified functions can be obtained by applying filters to the wavelet coefficients. In this way, we can process the shape geometry and synthesis new shapes to achieve different purposes, for example smoothing and deformation. This book also introduce some middle-level and high-level applications in Chapter 12, including shape matching, shape registration, and shape retrieval. These applications are built upon the low-level tools of shape representation and feature detection. They further demonstrate the great impact of wavelet tools in computer graphic and computer vision.

As a reference in computer graphics, this book includes implementations and algorithms too. In Chapter 8, we present key implementation details for the aforementioned diffusion-driven wavelets on discrete shapes. We give different schemes for implementing discrete Laplace-Beltrami operators on meshes and point clouds. One of them utilizes a previous result, which states that the short-time convergence (i.e. initial condition) of the Mexican hat wavelet is the Laplace-Beltrami operator. The eigen-system of the Laplace-Beltrami operator can be obtained by solving the generalized eigenvalue problem. The heat kernel and the Mexican hat wavelet are implemented based on a small number of Laplace-Beltrami eigenfunctions. We also present the implementation of matrix multiplication, which is used by the admissible diffusion wavelets and the anisotropic diffusion wavelets.

Not only presenting existing results, this book also opens a dialog for wavelet generation in Chapter 7. It extends the aforementioned diffusion-driven wavelets in two directions. One is to construct wavelets on complex data, i.e. volumes with multiple materials. We use the work in [Li et al. 12] to show the design of anisotropic wavelet on volumetric images. It is achieved by a diffusion tensor space built on the eigen-system of a local Hessian matrix. The other direction, inspired by the work in spectral graph wavelet [Antoine et al. 10,Hammond et al. 11], is to generate wavelets in the Fourier domain by wavelet generators. The special idea behind the scene is from the design of the Mexican hat wavelet, that functions without closed-form expressions on manifolds that may have analytical expressions in Fourier domain. By the manifold harmonics, it is feasible to design functions in frequency domain. We give some prototypes of some well-known wavelets in signal processing, which can be studied by interested readers.

The goal of this book is to give a thorough introduction of diffusion-driven wavelet design on manifold geometry, from exemplar design methods, generation schemes, to state-of-the-art applications in shape analysis. This book is valuable to researchers in computer graphics related areas, by means of collecting recent research results in wavelet design and application. These results made significant contributions, which facilitate space-frequency analysis of wavelet transform on manifold geometry. Besides, some unexplored topics are prompted to inspire new research directions. The methods and application in the book are valuable for research in computer vision, visualization, medical imaging, and geometric modeling as well. This book is also valuable to graduate students and senior undergraduate students majored in computer science. It explains profound theories by easy-understanding words and simple examples. Implementation details and application algorithms are also given, so that students are able to reproduce and use them for their studies.

Part I

Theories

2

Wavelet Theory

This chapter introduces basic knowledge of the wavelet theory that is related to this book. Starting from classical wavelets on Euclidean domains, it turns to wavelet design on structure data: graphs, meshes, and manifolds.

2.1 Classical Wavelet

As by its name, a wavelet is a function that oscillates like a wave along its domain, with attenuated amplitude that makes it locally-supported by a bounded region. It is much easier to get a first impression on wavelet by looking at some well-known examples in Fig. 2.1. Rather than theorems and equations, we would like to use three basic features for the first impression of wavelet: oscillation, attenuation, and multiscale.

2.1.1 Continuous Wavelet

The wavelet theory can be tracked back to the early 1980s by [Morlet et al. 82a, Morlet et al. 82b]. It has been applied to many engineering fields [Daubechies 92, Meyer 92, Meyer 01, Chan and Shen 05, Mallat 08]. When born, wavelets were designed for signal processing in some engineering fields with time-frequency localization. To understand this, we will need to pull the origin back to the Fourier transform.

For an integrable function $f(x)$, its Fourier transform is given by

$$\widehat{f}(\omega) = \int_{-\infty}^{\infty} f(x)e^{-i\omega x}dx, \tag{2.1}$$

where $e^{-i\omega x}$ is the Fourier basis, and ω is the frequency. The Fourier basis comprises a series of orthonormal functions with different periods, as shown in Fig. 2.2. The Fourier transform is to project functions to basis functions at different frequencies. The distribution of projected components makes another perspective for analyzing functions.

While in many references of wavelet, [Daubechies 92, Jaffard et al. 01, Mallat 08, Meyer 92], functions live in the time domain, we refer to the

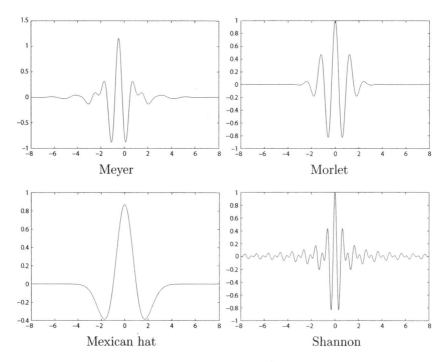

Figure 2.1. Wavelet examples: Meyer wavelet, Morlet wavelet, Mexican hat wavelet, and Shannon wavelet.

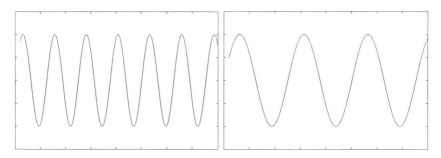

Figure 2.2. Two Fourier basis functions.

space domain in this book. The Fourier transform is a great tool for signal processing; however, it does not have localization in space. That is to say, it lacks the ability to extract contents locally in space.

One attempt to improve space localization is the short-time Fourier transform, which adds a window function to the Fourier transform

$$\widehat{f}(\omega, b) = \int_{-\infty}^{\infty} f(x)w(x - b)e^{-i\omega x}dx. \tag{2.2}$$

Although we consider space domain rather than time domain for Fourier transform and wavelet transform, we still adopt the name "short-time" for the consistency. The window function $w(x - y)$ therefore has non-zero values only in a small neighborhood centered at point y. It truncates the Fourier basis functions to fragments in space.

Now we can introduce wavelet. A family of wavelets $\psi_{a,b}(x)$ is generated from a single function $\psi(x)$ by dilation and translation,

$$\psi_{a,b}(x) = \frac{1}{\sqrt{|a|}} \psi\left(\frac{x - b}{a}\right). \tag{2.3}$$

The dilation parameter a and translation parameter b equip the wavelet with localization in both frequency and space. The normalization factor $\frac{1}{\sqrt{|a|}}$ ensures that $\|\psi_{a,b}(x)\|$ is independent of a and b.

The function $\psi(x)$ is often called the *mother* wavelet. The mother wavelet is a function that is oscillated and localized, which means it has to be bounded in a compact region. The mother wavelet is often assumed to satisfy the admissibility condition,

$$C_\psi = \int_{\mathbb{R}} \frac{|\widehat{\psi}(\omega)|^2}{|\omega|} d\omega < \infty, \tag{2.4}$$

where $\widehat{\psi}(\omega)$ is the Fourier transform of $\psi(x)$. The admissible condition implies

$$\widehat{\psi}(0) = \int \psi(x) dx = 0, \tag{2.5}$$

so the wavelet must be zero-mean.

For a given function integrable $f(x)$, the wavelet transform at scale a and location b is given by the inner product

$$\mathcal{W}_f(a, b) = \int_{\mathbb{R}} \psi_{a,b}(x) f(x) dx. \tag{2.6}$$

The wavelet is usually defined with unit energy

$$\int |\psi(x)|^2 dx = 1. \tag{2.7}$$

Therefore, the $\mathcal{W}_f(a, b)$ preserves the energy of function $f(x)$ after transform. The inverse wavelet transform is given by

$$f(x) = \frac{1}{C_\psi} \int_{\mathbb{R}^2} \mathcal{W}_f(a, b) \psi_{a,b}(x) \frac{da\, db}{a^2}. \tag{2.8}$$

2.1.2 Discrete Wavelet

In many applications, people are interested in the discrete wavelet transform, working with discrete functions. A discrete wavelet may be given by

$$\psi_{m,n}(x) = a^{-m/2}\psi(a^{-m}x - bn). \tag{2.9}$$

It should be noted that parameters a and b, indicating speeds of dilatation and translation, are different with the ones in the continuous wavelet. From now on, we will select a typical usage of $a = 2$ and $b = 1$, for the ease of understanding. Let us rewrite the discrete wavelet

$$\psi_{m,n}(x) = 2^{-m/2}\psi(2^{-m}x - n), \tag{2.10}$$

where variables m and n embody localization in frequency and space.

The construction of discrete wavelets is found upon a multiresolution analysis. Through a scaling function

$$\varphi_{m,n}(x) = 2^{-m/2}\varphi(2^{-m}x - n), \tag{2.11}$$

we obtain a chain of nested subspaces of the Hilbert space of integrable functions

$$V_m \subset V_{m-1}\cdots \subset V_1 \subset V_0 = L^2.$$

The wavelet space W_m, spanned on wavelets $\psi_{m,n}$ is the orthogonal complement of scaling space V_m inside V_{m-1}

$$V_{m-1} = V_m \oplus^{\perp} W_m.$$

Fig. 2.3 shows an example of discrete wavelet: the Haar wavelet. The scaling function φ_m is a smoothing function. As m increases, the scaling function has a larger supporting area, indicating a coarser representation of the function space. Therefore, the space can be compressed through the nested chain. The wavelet is simply the difference between two scaling functions:

$$\psi_m = \varphi_{m-1} - \varphi_m. \tag{2.12}$$

The discrete wavelet transform of a function $f(x)$ is given by the inner product

$$\mathcal{W}_f(m,n) = \langle \psi_{m,n}(x), f(x) \rangle. \tag{2.13}$$

Considering the first level scaling function $\varphi_{1,n}$ and wavelet function $\psi_{1,n}$ as low-pass filter g_n and high-pass filter h_n, the function is decomposed by repeatedly applying the two filters, which is known as a filter bank.

We will not go into any more details of classical wavelet, as the overview above is sufficient to understand the designed wavelets in this book.

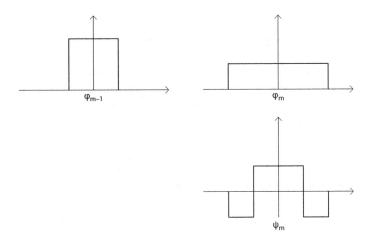

Figure 2.3. Discrete wavelet example: Haar wavelet.

2.2 Subdivision Wavelet

Building wavelets on curved surfaces is not an easy task in computer graphics. The early attempts are mostly based on subdivision surface.

2.2.1 Subdivision Scheme

A subdivision is a refinement scheme applied to a coarse piecewise linear polygon mesh, which can be used for representing a smooth surface. The subdivision scheme recursively subdivides each face into smaller faces. Fig. 2.4 shows two very simple examples of subdivision scheme: Catmull-Clark [Catmull and Clark 78] and Loop [Loop 87].

Fig. 2.5 shows a few steps of loop subdivision. A simple mesh of two tetrahedrons is gradually refined to a dense mesh close to a sphere. Subdivision provides an efficient way to represent continuous surfaces by a coarse control mesh and a refinement scheme.

2.2.2 Subdivision Wavelet

A subdivision scheme can be expressed by a matrix S, which subdivides a coarser mesh M_j to a finer mesh M_{j+1}:

$$M_{j+1} = SM_j. \tag{2.14}$$

The subdivision refinement is from coarse to fine. Recall that the nested subspace in discrete wavelet construction is from fine to coarse. One may come up with the idea that the reverse process of subdivision can be used

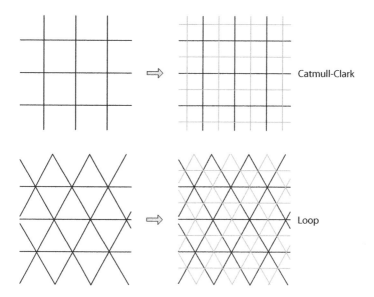

Figure 2.4. Subdivision scheme examples: Catmull-Clark (Top) and Loop (Bottom). Shaded lines are edges connecting newly inserted vertices by the subdivision schemes.

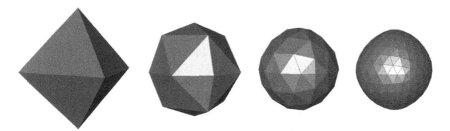

Figure 2.5. Loop subdivision surface.

for constructing wavelets on curved surfaces. As shown in Fig. 2.6, from a fine mesh M_{j+1}, by a reverse subdivision S^{-1} we obtain a coarser mesh M_j and wavelet coefficient W_j.

The subdivision wavelet was originally proposed by Lounsbery et al. [Lounsbery et al. 97]. The subdivision scheme that iteratively refines the mesh geometry also refines the functions. The constructed wavelets are biorthogonal and locally supported. The subdivision wavelets rely on the subdivision connectivity of the mesh, which restricts the application scope to data compression and level-of-detail rendering. Guskov et al. [Guskov et al. 99] first generalized basic signal processing tools to triangular meshes

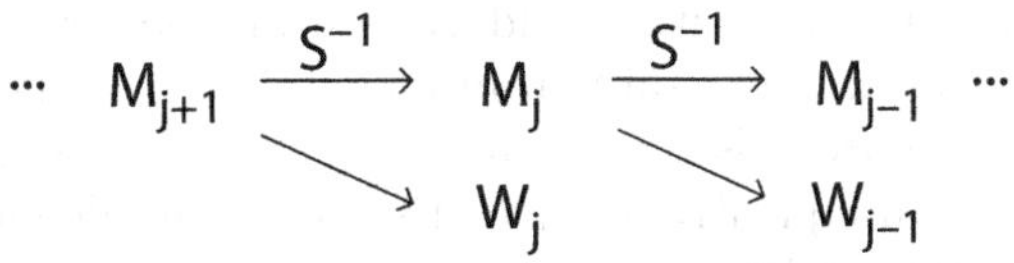

$$\cdots \; M_{j+1} \; \xrightarrow{\; S^{-1} \;} \; M_j \; \xrightarrow{\; S^{-1} \;} \; M_{j-1} \; \cdots$$
$$\searrow \; W_j \qquad \searrow \; W_{j-1}$$

Figure 2.6. Subdivision wavelet.

with irregular connectivity, through subdivision wavelets. Daubechies et al. [Daubechies et al. 99] studied subdivision wavelets built on irregular point sets. In [Bertram et al. 00], Bertram et al. utilized bicubic B-spline subdivision to construct wavelet transform that affords boundary curves and sharp features. In [Bertram et al. 04], B-spline wavelets were combined with the lifting scheme for biorthogonal wavelet construction. As a drawback, the subdivision wavelet requires the meshes to have subdivision connectivity, where remeshing process is frequently needed. To avoid remeshing, Valette and Prost [Valette and Prost 04] extended the subdivision wavelet for triangular meshes using irregular subdivision scheme that can be directly computed on irregular meshes. On spherical domains, Haar wavelets [Bonneau 99, Nielson et al. 97] were constructed over nested triangular grids generated by subdivision. Recently, the spherical Haar wavelet basis was improved to both orthogonal and symmetric by [Lessig and Fiume 08]. In subdivision wavelets, the dilation of scaling functions strictly follows the subdivision scheme, which depends on the meshing. In [Wang et al. 07], a biorthogonal wavelet analysis based on the $\sqrt{3}$-subdivision was proposed. It is a well orchestrated solution on triangular meshes since the $\sqrt{3}$-subdivision is of the slowest topological refinement among all the traditional triangular subdivisions. In a recent work [Charina et al. 10], Charina et al. constructed compactly supported tight frames of multivariate multiwavelets with subdivision schemes.

The essential motivation is to apply the MRA to arbitrary surfaces. The subdivision seeks to model a smooth surface via a recursive process of refining polygonal faces from a coarse base mesh, which is essentially a model-driven, top-down methodology towards wavelets definition and construction. The subdivision wavelets have been frequently used for geometry compression and level-of-detail data visualization, such as volume rendering [Lippert and Gross 95], scientific visualization [Craciun et al. 05], spectral rendering [de Iehl and Péroche 00], multiresolution for surfaces [Olsen et al. 07], and animation compression [Payan and Antonini 07].

Subdivision wavelet is actually a by-product of subdivision surface. It changes the shape, and therefore the domain. The construction requires the subdivision hierarchy before defining wavelets, which also limits its application scope. Nowadays, in conjunction with the rapid technology advancement of data acquisition and computing power, many graphics problems call for a paradigm shift from model-driven to data-driven. The regularly-refined hierarchy is computationally expensive and perhaps even harder to build. Consequently, it gives rise to strong demand in flexibly adapted wavelet tools without building the subdivision explicitly, which can be used for fast space-frequency processing. Applications (that are critically enabled) span traditional geometry processing to visual analytics, feature extraction, feature-driven data mapping, etc., many of which require local operations on fine details at different frequencies.

2.3   Diffusion Wavelet

Recently, a new methodology to construct diffusion wavelets [Coifman and Maggioni 06] on graphs and manifolds has been proposed. In sharp contrast to the subdivision wavelets, the diffusion wavelets are built by a fundamentally-different, bottom-up philosophy that starts from the fine input data. This new method inspires us in alternative directions for building discrete wavelets on curved surfaces.

2.3.1   Diffusion Wavelets

Diffusion wavelets are a set of bottom-up built wavelets based on diffusion. The construction of diffusion wavelets adopts an operator T and its powers to expand the nested subspaces, where scaling functions and wavelets are obtained by orthogonalization and rank-revealing compression. Examples of diffusion operator were given as $T = e^{-\epsilon \mathcal{L}}$ and $T = I - \epsilon \mathcal{L}$, where ϵ is a small positive parameter. Operator T should be self-adjoint, with spectrum

$$1 = \lambda_0 \geq \lambda_1 \geq \lambda_2 \geq \ldots .$$

It starts from the basis $\Phi_0 = \{\delta_x\}$ of the initial space $V_0 = L^2$. The columns of T can be interpreted as the set of functions $\{T\delta_x\}$. The basis $\Phi_1 = \{\varphi_{1,x}\}$ is obtained by a local multiscale orthogonalization procedure. Since the subspace V_1 is smoothed once by the operator T, Φ_1 has a lower rank than Φ_0. That is to say, the subspace is downsampled.

The powers of the operator T decrease in rank, and therefore, compress the geometric space spanning at each power. The powers dilate the operator T to a set of discrete scales: $j = 1, 2, \ldots$. At scale j, the function space is dilated to T^{2^j}, which is further compressed to a subspace V_j by the sparse QR decomposition according to its numerical rank. A chain of nested subspaces form a multiresolution analysis:

$$V_j \subset V_{j-1} \ldots \subset V_1 \subset V_0 = L^2 .$$

Algorithm 1: Diffusion wavelet tree.

Input: operator matrix T, precision of the QR decomposition ϵ,
number of scale levels J

Output: scaling basis $\{\Phi_j\}$, wavelet basis $\{\Psi_j\}$

for $j = 0 : J - 1$ **do**

$\quad [\Phi_{j+1}]_{\Phi_j}, [T]_{\Phi_0}^{\Phi_1} \leftarrow QR([T^{2^j}]_{\Phi_j}^{\Phi_j}, \epsilon);$

$\quad [T^{2^{j+1}}]_{\Phi_{j+1}}^{\Phi_{j+1}} \leftarrow [\Phi_{j+1}]_{\Phi_j} [T^{2^j}]_{\Phi_j}^{\Phi_j} [\Phi_{j+1}]_{\Phi_j}^{*};$

$\quad [\Psi_j]_{\Phi_j} \leftarrow QR(I_{\langle \Phi_j \rangle} - [\Phi_{j+1}]_{\Phi_j} [\Phi_{j+1}]_{\Phi_j}^{*}, \epsilon);$

end

The orthogonal complementary space between two adjacent subspaces is defined as the wavelet space W_j, such that

$$V_{j-1} = V_j \oplus^{\perp} W_j.$$

An orthonormal basis Φ_j of scaling functions is obtained by the sparse QR decomposition from the subspace V_j, while another orthonormal basis Ψ_j of wavelets is obtained from the complementary space W_j. The algorithm for computing diffusion wavelets is given by Algorithm 1. For more details, please refer to [Coifman and Maggioni 06].

Fig. 2.7 shows some scaling functions and wavelets at the same level. The functions are obtained by online source code [1], used in [Coifman and Maggioni 06]. Diffusion wavelets are not well localized. For basis functions at the same level, their supported regions have different sizes that can be as large as the entire domain. When the operator is a diffusion type operator, the diffusion wavelets model a diffusion process. This diffusion-driven methodology naturally dilates the functions associated with the underlying heat diffusion process, which solely depends on data geometry.

We are not going any deeper in the theory of diffusion wavelets, but would like to introduce some applications. In [Maggioni 07] and [Wang and Mahadevan 09], diffusion wavelets were adopted for document corpora analysis. Diffusion wavelets can build multiscale embedding of the documents. This multiscale representation significantly compresses the data at different scales, while preserving important information. As a wavelet tool, it also extracts details of the documents of different scales. The document corpora analysis can be used for document classification, clustering, and retrieval. In [Mahadevan and Maggioni 05], diffusion wavelets were used for value function approximation. In particular, two approaches were proposed using eigenfunctions of the Laplacian and the diffusion wavelets. The eigenfunctions of graph Laplacian are like Fourier basis functions, which will be

---

[1] http://www.math.duke.edu/~mauro/code.html

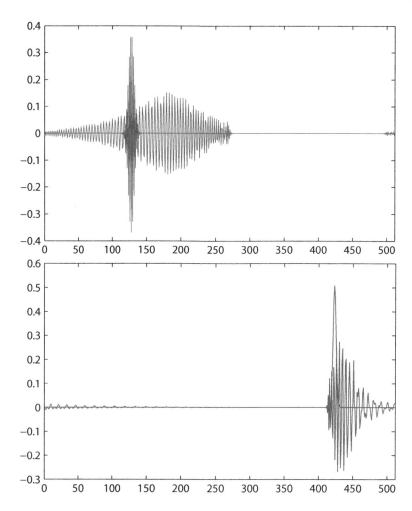

Figure 2.7. Some scaling functions and wavelets of diffusion wavelets. The scaling functions and wavelets are from the same level.

discussed more in the next section. The approach using diffusion wavelets builds random walks on the graph, which delivers a multiscale representation for function approximation. Interestingly, the two methods are both utilized for constructing wavelets on 3D shapes in this book.

2.3.2 Biorthogonal Diffusion Wavelets

As an improvement, the biorthogonal diffusion wavelets [Maggioni et al. 05] were proposed, relieving the excessively-strict orthogonality property of scaling functions. This is achieved by using a Riesz basis Φ_j instead of an orthonormal basis, and introducing a dual Riesz basis $\widetilde{\Phi}_j$, such that

$$\widetilde{\Phi}_j^* \Phi_j = I.$$

Hence for any function $f \in V_j$, we have

$$f = \sum_k \langle f, \widetilde{\varphi}_{j,k} \rangle \varphi_{j,k}, \tag{2.15}$$

which is a biorthogonal reconstruction.

A Riesz basis r_k, in a nutshell, is a basis that can be obtained from an orthogonal basis e_k and an invertible transformation P, such that

$$r_k = P e_k, \text{ for all } k.$$

A Riesz basis is independent, but not necessarily orthogonal. For a Riesz basis r_k, there exists a dual basis $\widetilde{r}_k$ that is also a Riesz basis.

Therefore, basis Φ_j is not necessarily orthogonal, or more precisely, it is biorthogonal. The same procedure is applied to the wavelet basis Ψ_j as well, which now has a biorthogonal dual $\widetilde{\Psi}_j$. Because of this change, the scaling functions and wavelets can be locally-supported. Since they are not orthogonal, their duals are needed for reconstruction.

Fig. 2.8 shows some scaling functions of biorthogonal diffusion wavelets at two levels. Biorthogonal diffusion wavelets have better localization than diffusion wavelets. The scaling functions are smooth functions. By dropping the orthonormalization, the biorthgonal diffusion wavelets allow more flexible constructions than the diffusion wavelets.

2.4 Spectral Graph Wavelet

In mathematics, researchers have found a very neat construction of continuous wavelet of graphs. The construction is based on the spectral graph theory, for example the work in [Hammond et al. 11].

2.4.1 Spectral Graph Theory

Graph is a nice representation for structure data, relation data, manifold data, etc. The spectral graph theory studies eigenvalues and eigenvectors of matrices associated to a graph. It helps people know the graph as spectrum, analogous to the Fourier transform. It is a fundamental property of graphs.

For a weighted graph G with weights $w_{i,j}$, the graph Laplacian $\mathcal{L}$ is a matrix representation of the graph. The Laplacian matrix can be computed

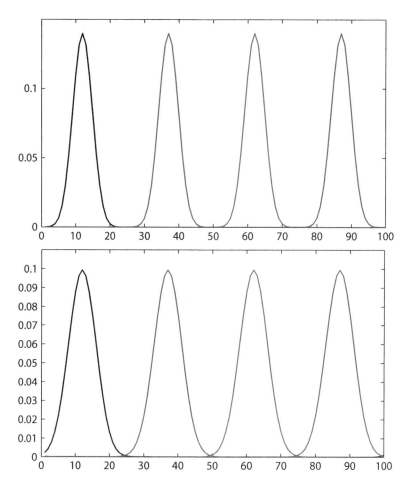

Figure 2.8. Some scaling functions of the biorthogonal diffusion wavelets at two levels.

by the difference of the degree matrix D and the adjacency matrix A:

$$\mathcal{L} = D - A. \tag{2.16}$$

The diagonal matrix D is the degree matrix with

$$D_{i,i} = \sum_{j \in N(i)} w_{i,j}, \tag{2.17}$$

where $N(i)$ is the neighborhood of node i, and $w_{i,j}$ is the weight of edge

(i, j). Hence, the Laplacian matrix $\mathcal{L}$ has entries

$$\mathcal{L}_{i,j} = \begin{cases} \sum_{k \in N(i)} w_{i,k}, & i = j \\ -w_{i,j}, & \text{otherwise} \end{cases}. \tag{2.18}$$

The normalized Laplacian matrix $\mathcal{L}^{norm}$ is defined as

$$\mathcal{L}^{norm} = I - D^{-1/2} A D^{-1/2} = D^{-1/2} L D^{-1/2}. \tag{2.19}$$

It has entries

$$\mathcal{L}_{i,j}^{norm} = \begin{cases} 1, & i = j \\ -\dfrac{1}{\sqrt{d_{i,i} d_{j,j}}}, & \text{otherwise} \end{cases}. \tag{2.20}$$

Usually the normalized Laplacian matrix is used in computation instead of the Laplacian matrix. For simplicity, we will just say the Laplacian matrix.

It is well-known that the Laplacian matrix is symmetric. It has real non-negative eigenvalues

$$0 = \lambda_0 \le \lambda_1 \le \cdots \le \lambda_{n-1},$$

which is called the spectrum of the graph. Its eigen-system $\{\lambda_k, \phi_k\}_{k=0}^{n-1}$ has

$$\mathcal{L}\phi_k(x) = \lambda_k \phi_k(x). \tag{2.21}$$

The eigenvectors have very similar properties with the Fourier basis functions, which are both periodic, orthogonal, and complete. The set of eigenvectors compose a complete basis for the function space of the graph.

The graph Fourier transform of a function f on graph is given by

$$\widehat{f}(k) = \langle f(x), \phi_k(x) \rangle, \tag{2.22}$$

and the inverse transform is given by

$$f(x) = \sum_{k=0}^{n-1} \widehat{f}(k) \phi_k(x). \tag{2.23}$$

2.4.2 Spectral Graph Wavelet

The spectral graph wavelets are generated by the choice of a wavelet kernel

$$g : \mathbb{R}^+ \to \mathbb{R}^+.$$

Wavelet kernels are analogous to the classical wavelet in Fourier domain: $\widehat{\psi}(k)$. They should behave as band-pass filters, for example the Mexican hat wavelet in Fourier domain (right figure in Fig.5.1).

Based on the spectral graph theory, Hammond et al. designed spectral graph wavelets by deriving a wavelet operator from the graph Laplacian: $T_g = g(\mathcal{L})$. The Fourier transform of this operator is given by

$$\widehat{T_g}(k) = g(\lambda_k). \tag{2.24}$$

When applying this operator to a function f, we get the formulation in Fourier domain

$$\widehat{T_g f}(k) = g(\lambda_k)\widehat{f}(k). \tag{2.25}$$

By the inverse Fourier transform, we have

$$(T_g f)(x) = \sum_{k=0}^{n-1} g(\lambda_k)\widehat{f}(k)\phi_k(x). \tag{2.26}$$

The wavelet operator is dilated by time t. At time t, the wavelet operator is given by

$$T_g^t = g(t\mathcal{L}). \tag{2.27}$$

The scaling is continuous in t. This yields a continuous wavelet generation.

The wavelet operator can be interpreted as a wavelet continuous in space. The wavelets are realized through localizing the wavelet operator by the impulse function,

$$\psi_t(x,y) = T_g^t(x,y)\delta(x,y). \tag{2.28}$$

with expansions on the graph Laplacian eigen-decomposition

$$\psi_t(x,y) = \sum_{k=0}^{N-1} g(t\lambda_k)\phi_k(x)\phi_k(y). \tag{2.29}$$

For a function $f(x)$ on a graph, the spectral graph wavelet transform is defined as the inner product

$$\mathcal{W}_f(x,t) = \langle \psi_t(x,y), f(y) \rangle. \tag{2.30}$$

By the inverse graph Fourier transform, we get the following expansion as well

$$\mathcal{W}_f(x,t) = \sum_{k=0}^{N-1} g(t\lambda_k)\widehat{f}(k)\phi_k(x), \tag{2.31}$$

where $\widehat{f}$ is the graph Fourier transform of f.

In [Hammond et al. 11], Hammond et al. also showed the inverse transform of spectral graph wavelets, which is analogous to the inverse transform of classical wavelets. If the wavelet kernel g satisfies the admissibility condition

$$C_g = \int_0^\infty \frac{g^2(x)}{x} dx < \infty, \qquad (2.32)$$

and $g(0) = 0$, the function f can be reconstructed by the inverse transform

$$f(x) = \widehat{f}(0)\phi_0(x) + \frac{1}{C_g} \sum_{k=1}^{N-1} \int_0^\infty \mathcal{W}_f(x,t)\phi_k(x)\frac{dt}{t}, \qquad (2.33)$$

where $\widehat{f}(0)\phi_0(x)$ is a residual constant. We refer reader to [Hammond et al. 11] for a detailed proof.

In Chapter 5, we will show that for the Mexican hat wavelet, this inverse transform can be further simplified to a neat expression that only contains wavelet transforms.

In [Antoine et al. 10], Antoine et al. presented a very similar work for spectral graph theory. In this book, we will show that this approach can be applied to general manifold data, by replacing the graph Laplacian with the Laplace-Beltrami operator. This indicates the relation between wavelet design and heat diffusion from one aspect.

3

Heat Diffusion Theory

This chapter briefly introduces the heat diffusion theory, including heat equation and heat kernel. Heat diffusion models the physical phenomenon of heat transfer and distribution on a surface over time. It is related to partial differential equation (PDE), Brownian motion, and Fourier transform. When applied to different types of data, it has a wide range of applications in multiscale analysis, spectral analysis, probability theory, statistical learning, data mining, financial mathematics, social network, etc. Particularly in computer graphics, it naturally connects key subjects in stochastic differential geometry curvature, spectrum, Brownian motion, and topology.

We studied the heat diffusion theory as it is an effective method for wavelet design. A representative example is the Mexican hat wavelet, which is the negative first derivative of the Gaussian diffusion kernel, with respect to t.

3.1 Heat Equation

3.1.1 Definition

In a mathematically rigorous manner, we consider compact Riemannian manifolds, serving as domains for heat diffusion. Let (M, μ) be a n-dimensional compact Riemannian manifold with possible boundary ∂M. Its volume μ is defined by

$$d\mu = \sqrt{g}dx^1 dx^2 \dots dx^n, \tag{3.1}$$

where g is the metric of M, and $x^1, x^2, \dots, x^n$ are coordinates in the local chart. For ease of understanding, a 3D curved shape without self-interaction is a 2D manifold embedded in $\mathbb{R}^3$. Its volume is the total surface area.

It is well known that the heat diffusion process on manifolds is governed by the heat equation. For a square integrable function $u(x, t)$ at point $x \in M$ and time $t > 0$, the heat equation is usually written as

$$\frac{\partial u(x, t)}{\partial t} - \Delta_M u(x, t) = 0, \tag{3.2}$$

Here, Δ_M denotes the Laplace-Beltrami operator on (M, μ) given by

$$\Delta_M = \frac{1}{\sqrt{g}} \sum \frac{\partial}{\partial x^i} \left(g^{ij} \sqrt{g} \frac{\partial}{\partial x^j} \right), \qquad (3.3)$$

with $(g^{ij}) = (g_{ij})^{-1}$, and $g = \det(g_{ij})$.

It may be noted that in some literature, the minus sign in Eq.(3.2) is replaced by a plus sign, which changes nothing but signs of expressions derived from the heat equation. The diffusion equation is an important PDE, which describes density fluctuations in a material undergoing diffusion. In the physical problem of temperature variation, $u(x, t)$ is the temperature of point x at time t. The heat equation states that the rate of temperature change (heat inside) at a point balances to the flux passing in or out of it. In differential geometry, it was used to define the Ricci flow and solve the *Poincaré conjecture*.

3.1.2   Conditions

An *initial condition* of the heat equation is a solution at $t = 0$

$$u(x, 0) = u_0(x).$$

It specifies an initial heat distribution before diffusion. Heat diffusion is a smoothing process of this initial condition. Eventually, it will reach a *steady condition* that has

$$\frac{\partial u(x, t)}{\partial t} = 0,$$

which means the function $u(x, t)$ comes to a constant distribution that does not change over time. Meanwhile, because of

$$\Delta_M u(x, t) = 0,$$

$u(x, t)$ is constant everywhere on the manifold, or in another saying, heat is equally distributed on the domain. Also because the total heat is conserved in the heat equation over t, the steady condition can be computed by

$$u(x, t) = \frac{1}{\mu(M)} \int_M u_0(x) d\mu(x).$$

If the manifold has boundary $\partial M \neq \emptyset$, it needs some *boundary condition* to make it self-adjoint. Boundary conditions impose restraints on values and/or derivatives of boundary ∂M. Some common boundary conditions are listed here.

- Dirichlet (or first-type) condition: for $x \in \partial M$

$$u(x, t) = f(x).$$

- Neumann (or second-type) condition: for $x \in \partial M$

$$\frac{\partial u(x,t)}{\partial \mathbf{n}} = g(x),$$

  where $\mathbf{n}$ denotes the exterior normal to the boundary ∂M.

- Robin (or third-type) condition: for $x \in \partial M$

$$au(x,t) + b\frac{\partial u(x,t)}{\partial \mathbf{n}} = h(x),$$

  where a and b are combination coefficients.

In this book, we will always use Neumann condition for shapes with boundaries. One of the reasons for using Neumann condition is explained below.

3.1.3 Eigen-System

The Laplace-Beltrami operator has well-defined eigen-system $\{\lambda_k, \phi_k\}_{k=0}^{\infty}$ for a compact manifold M:

$$\Delta_M \phi_k(x) = -\lambda_k \phi_k(x), \tag{3.4}$$

where λ_k and $\phi_k(x)$ are the k-th eigenvalue and the k-th eigenfunction of the Laplace-Beltrami operator, respectively. The spectrum of Δ_M consists of an increasing positive sequence

$$0 \le \lambda_0 < \lambda_1 < \cdots < \lambda_\infty.$$

Since the Laplace-Beltrami operator Δ_M characters a manifold M, this spectrum is called fingerprint [Reuter et al. 05] or DNA [Reuter et al. 06] of a shape. The eigenfunctions $\{\phi_k\}_{k=0}^{\infty}$ form an orthonormal and complete basis for the Hilbert space $L^2(M)$. For manifolds without boundaries and with Neumann-conditioned boundaries, we have the first eigenvalue $\lambda_0 = 0$, and the first eigenfunction is constant everywhere on the manifold

$$\phi_0(x) = \frac{1}{\sqrt{\mu(M)}}.$$

At boundary $x \in \partial M$, we have

$$\frac{\partial \phi_i(x)}{\partial \mathbf{n}} = 0,$$

and thus we do not need to set fixed values for eigenfunctions at boundary.

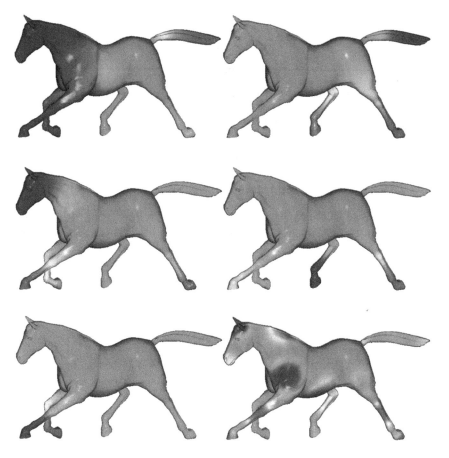

Figure 3.1. The 1st, 2nd, 3rd, 4th, 5th, and 10th Laplace-Beltrami eigenfunctions of a closed shape, rendered by coded color. See color insert.

Fig. 3.1 shows some Laplace-Beltrami eigenfunctions of a complete shape, rendered by coded color. Eigenfunctions are functions with periodic values over the domain. Fig. 3.2 shows the first 300 eigenvalues of the horse model. Its asymptote is close to a straight line.

The Laplace-Beltrami eigen-system is analogous to Fourier frequency and basis. The traditional Fourier transform of an integrable function $f(x)$ in $\mathbb{R}^n$ is defined as

$$\widehat{f}(\omega) = \int_{-\infty}^{\infty} f(x)e^{-i\omega x}dx, \tag{3.5}$$

where $e^{-i\omega x}$ is the Fourier basis, and ω is the frequency. By Euler's formula, the Fourier basis ends up with sine and cosine functions, which are

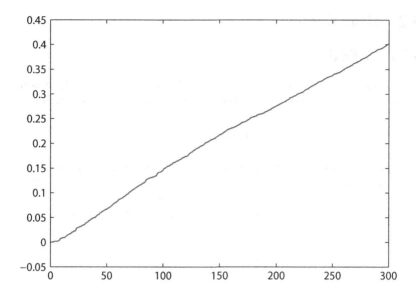

Figure 3.2. The first 300 eigenvalues of the horse model. Its asymptote is close to a straight line.

orthonormal and oscillating over the domain. They can be interpreted as eigenfunctions of derivative operators with respect to x. For example, $e^{-i\omega x}$ is the eigenfunction of the second-order derivative operator associated with the eigenvalue ω^2, because of

$$\frac{d^2}{dx^2}e^{-i\omega x} = \omega^2 e^{-i\omega x}. \tag{3.6}$$

That is, by taking the Fourier basis as eigenfunctions of the second-order derivative operator, the eigenvalues are real numbers corresponding to the squares of frequencies. Similarly, we can define the Fourier transform of a function $f(x) \in L^2(M)$ on manifold M,

$$\widehat{f}(k) = \langle f(x), \phi_k(x) \rangle. \tag{3.7}$$

With a slight abuse of language, we call $\widehat{f}(k)$ the Fourier transform, or Fourier coefficient, of function $f(x)$ on manifold M. In [Vallet and Lévy 08], it is also called the manifold harmonics transform.

3.2   Heat Kernel

3.2.1   Definition

Heat kernel is a fundamental solution to the heat equation, which is used to solve PDEs in mathematics. Assume we set the initial condition as

$$u_0(x) = \delta(x),$$

a Dirac delta function, which means we put a finite amount of infinitely hot heat source at the origin and the instant $t = 0$. Heat then diffuses to everywhere on the manifold. The temperature remains highest at the origin, but the heat is scatted over the manifold. The distribution of heat forms a fundamental solution, i.e. a heat kernel

$$h_t(x,y) : \mathbb{R}^+ \times M \times M \to \mathbb{R}^+.$$

It describes the amount of heat diffused from point x to point y at time t. The heat kernel can compute solutions to the heat equation with some initial condition $u_0(x)$, given by

$$u(x,t) = \int_M h_t(x,y)u_0(y)d\mu(y), \tag{3.8}$$

where $\mu(y)$ denotes the Riemannian volume of y on the manifold.

When considering d-dimensional Euclidean space $\mathbb{R}^d$, the heat kernel has an explicit expression that is a Gaussian

$$h_t(x,y) : G_t(x,y) = \frac{1}{(4\pi t)^{d/2}} e^{-\frac{\|x-y\|^2}{4t}}. \tag{3.9}$$

Gaussian function has been widely used in computer vision and computer graphics. It is the kernel to build multiscale analysis of images.

On manifolds, the solution of the heat kernel can be computed through the eigen-system of the Laplace-Beltrami operator. Assume a solution to the heat equation is square integrable: $u(x,t) \in L^2 M$. It can be expanded on the Laplace-Beltrami eigenfunctions:

$$u(x,t) = \sum_{k=0}^{\infty} c_k(t)\phi_k(x), \tag{3.10}$$

where $c_k(t)$ is the coefficient at the basis function $\phi_k(x)$. According to

[Strang 86], we have

$$\frac{\partial u(x,t)}{\partial t} = \sum_{k=0}^{\infty} \frac{\partial c_k(t)}{\partial t} \phi_k(x)$$

$$= \sum_{k=0}^{\infty} c_k(t) \Delta_M \phi_k(x) \tag{3.11}$$

$$= \sum_{k=0}^{\infty} -c_k(t) \lambda_k \phi_k(x).$$

Hence, we have

$$\frac{\partial c_k(t)}{\partial t} = -c_k(t) \lambda_k. \tag{3.12}$$

Combing the initial condition of the heat equation, we have

$$c_k(t) = c_k(0) e^{-\lambda_k t}, \tag{3.13}$$

where $c_k(0)$ is associated with $u_0(x)$. For the heat kernel $h_t(x,y)$, the initial condition is $u_0(x,y) = \delta(x,y)$. Since the eigenfunctions are orthonormal, it indicates

$$\delta(x,y) = \sum_{k=0}^{\infty} \phi_k(x) \phi_k(y). \tag{3.14}$$

Therefore, we obtain an expansion of the heat kernel, given by

$$h_t(x,y) = \sum_{k=0}^{\infty} e^{-\lambda_k t} \phi_k(x) \phi_k(y). \tag{3.15}$$

Formally differentiating the series under the sign of the summation shows that this should satisfy the heat equation and the initial Dirac condition.

3.2.2 Properties

Heat kernel has many nice properties that make itself a powerful tool in shape analysis. It is positive and symmetric:

$$h_t(x,y) > 0, \ h_t(x,y) = h_t(y,x).$$

It is isometric invariant, since the Laplace-Beltrami operator and its eigensystem are invariant to isometric deformation. Therefore, the heat kernel is intrinsic and characterizes the shape. This builds the foundation for the vast applications of heat kernel in processing dynamic shapes with isometric deformation.

Heat kernel satisfies the semigroup identity

$$h_{t+s}(x,y) = \int_M h_t(x,z)h_s(y,z)d\mu(z). \tag{3.16}$$

That is, a heat kernel can be accumulated from previous heat kernels. For example, if we have heat kernel $h_t(x,y)$ at t, we can get heat kernel $h_{2t}(x,y)$ at $2t$ by applying $h_t(x,y)$ twice, and so on. In another saying, we can use current heat kernel to infer future heat kernels. This is quite useful in computing heat kernels.

The heat diffusion is energy-conservative, hence the integration of heat kernels at one point and one time should be conservative

$$\int_M h_t(x,y)d\mu(y) = 1.$$

If we consider heat kernels from one point to itself all over the manifold, this integration turns to the heat trace

$$\int_M h_t(x,x)d\mu(x) = \sum_{k=0}^{\infty} e^{-\lambda_k t}. \tag{3.17}$$

The short-time convergence of heat kernel is the Dirac function

$$h_0(x,y) = \delta_x(y).$$

Meanwhile, the long-time converge of heat kernel is the steady condition

$$h_\infty(x,y) = \frac{1}{\mu(M)}.$$

Heat kernel is multiscale, with variable t corresponds to the scale. Small values of t correspond to small scales and thus high frequencies, while large values correspond to large scales with low frequencies. For a given time t, the majority of heat kernel $h_t(x,\cdot)$ falls in a local area around x. This is because the heat kernel has Gaussian decay over manifolds, and it is localized.

3.2.3   Interpretations

For a Brownian motion X_t, there is an increasing function $R(t)$ called an "upper radius" [Grigor'yan 99] if, with probability 1, we have $X_t \in B_{R(t)}(x)$ for all t large enough, where $B_r(x)$ denotes a geodesic ball centered at x with radius r. More precisely, the upper radius gives rise to a theoretic result given in the following theorem.

Theorem 3.1 (The Law of The Single Logarithm [Grigor'yan 99]) *Let M be a geodesically complete manifold. Assume that, for some $x \in M$ and all r large enough, the volume*

$$V_r(x) \leq \text{const } r^d,$$

with some $d > 0$. Then for any $\epsilon > 0$, the function

$$R(t) = \sqrt{(d + \epsilon)t \log t}$$

is an upper radius for the process X_t started at any $x \in M$.

This theorem can be used to find an approximate relation between time t and radius of the local supporting area. As time increases, this area is extended. For a manifold with bounded volume, it will eventually span the entire domain.

As a fundamental solution, the heat kernel is used for solving the heat equation. For the initial condition $u(0, x) = u_0(x)$, solutions to the heat equation are usually expressed by $H_t u_0(x)$, where H_t is the heat operator. Applying the heat operator to a function $f(x) \in L^2(M)$, equals to the integration

$$H_t f(x) = \int_M h_t(x, y) f(y) d\mu(y), \tag{3.18}$$

for all $t > 0$. Therefore, we can solve the PDE of heat diffusion by the heat kernel.

From the aspect of probability, the heat kernel can be interpreted as the transition density function of Brownian motion on the manifold. For a Borel subset $C \subseteq M$, the probability of Brownian motion X_t^x starting at x in C at time t will be

$$\int_C h_t(x, y) d\mu(y).$$

Brownian motion is the most basic continuous time Markov process on manifolds and graphs. The probabilistic interpretation well explains the stableness of heat kernel against noise and holes. That is, small perturbations and missing parts do not affect much the connectivity of two points, and hence the probability of Brownian motion.

From the aspect of geometry, the heat kernel has a short-time convergence [Grigor'yan 06], given by

$$\lim_{t \to 0} t \log h_t(x, y) = -\frac{1}{4} d_g^2(x, y), \tag{3.19}$$

where $d_g(x, y)$ is the geodesic distance between x and y on manifold M. Therefore, the heat kernel provides some geometric measure of points on manifolds. More importantly, it derives the diffusion distance $d_t(x, y)$, defined through

$$d_t^2(x, y) = h_t(x, x) + h_t(y, y) - 2h_t(x, y)$$
$$= \sum_{k=0}^{\infty} e^{-\lambda_k t} (\phi_k(x) - \phi_k(y))^2, \qquad (3.20)$$

which is exactly a metric. More related metrics for graphs or manifolds include commute-time distance [Fouss et al. 07], resistance distance [Wu 04], and biharmonic distance [Lipman et al. 10].

Heat kernel contains all the information about the intrinsic geometry of the manifold, and hence captures shapes up to isometry. This property was named "informative" in [Sun et al. 09].

For shape description, the heat kernel characterizes the shape up to isometry. In [Sun et al. 09], a concise representation was given by the heat kernel from one point to itself:

$$h_t(x, x) = \sum_{k=0}^{\infty} e^{-\lambda_k t} \phi_k^2(x), \qquad (3.21)$$

which is named Heat Kernel Signature (HKS). It forms the diagonal entries of the heat kernel matrix, which conserves all geometric information. Consequently, it is also informative. As an effective shape descriptor, it has been applied to dynamic shape analysis, such as feature [Sun et al. 09], matching [Ovsjanikov et al. 10, Dey et al. 10], registration [Hou and Qin 12b], and retrieval [Bronstein and Kokkinos 10, Bronstein et al. 11].

3.3   Applications in Shape Analysis

The diffusion theory has been applied to a wide range of applications in shape analysis, such as shape representation, multiscale approximation, feature detection, shape matching, segmentation, spectral analysis, geometry processing, and shape retrieval.

3.3.1   Heat Diffusion

Originally, heat diffusion is frequently used for building a multiscale representation of functions on data. Its simulation spontaneously generates a multiscale hierarchy of the initial condition. It prevails in image processing, where images are defined as functions on a 2D Euclidean domain. For 3D shapes, one can build a multiscale hierarchy of functions by conducting heat diffusion. In [Lee et al. 05], Lee et al. used a Gaussian kernel to conduct heat diffusion of curvature maps on 3D shapes:

$$f(x, \sigma) = \kappa(x) * G_\sigma(x, y), \qquad (3.22)$$

where $\kappa(x)$ denotes the curvature at vertex x, σ denotes the standard deviation of the Gaussian $G_\sigma(x, y)$, and $f(x, \sigma)$ consists a multiscale representation of the function $\kappa(x)$. Differences of adjacent scales in the multiscale representation record detailed information of shape geometry. This method was used for computing a saliency map of a shape, which is a non-linear combination of details at different scales. In [Zaharescu et al. 09], Zaharescu et al. repeatedly applied a fixed Gaussian to an input scalar field on meshes, based on the semi-group property of heat diffusion. The fixed Gaussian has a small standard deviation, which only affects a small area around the center point. This avoids computing expensive convolutions in large areas. The method was applied to feature detection and matching on deformable shapes.

In [Hou and Qin 10], we equipped the Gaussian kernel with geodesic metric, which is much closer to the heat kernel for large scales. For the computational aspect, we combined the kernel computation and semi-group convolution. That is, we compute kernels at differential scales, and boost convolutions from one scale to another. In [Hou and Qin 12a], we showed that accurate heat diffusion can be computed by convolution with the heat kernel. This convolution has a linear time complexity by spectral decomposition and Fourier transform. A similar work appears in [Patanè and Falcidieno 10], which builds a multiscale space by the heat kernel for shape processing and analysis. Beyond regular heat diffusion, we extended it to anisotropic diffusions. In [Wang et al. 11a], we proposed an anisotropic diffusion by a normal-driven shape representation. The isometric Laplace-Beltrami operator turns into an edge-weighted differential operator. It was applied to feature detection, mesh smoothing/denoising, and shape decomposition. In [Li et al. 12], we designed a diffusion-tensor weighted diffusion on volumetric images for registration. The diffusion tensor is formulated by means of Hessian matrix eigen-system that intrinsically encodes local geometric structure.

3.3.2 Laplace-Beltrami Eigen-System

Since the eigen-system of the Laplace-Beltrami operator in heat diffusion characterizes shape geometry, it has been used for shape processing. In [Lévy 06], Levy studies Laplace-Beltrami eigenfunctions $\{\phi_k(x)\}_{k=0}^\infty$ in geometry processing and pose transfer. The eigenfunctions compose a complete and orthogonal basis of the Hilbert space, which repeatedly oscillate on the manifold. As introduced before, the basis is analogous to the Fourier basis, and thus, there is an analogous Fourier transform on manifolds.

In [Vallet and Lévy 08], Vallet and Lévy proposed the manifold harmonic transform, which is an analogy of Fourier transform on manifolds. It facilitates signal processing of shape geometry in frequency domain. For a given function $f(x)$, its manifold harmonic transform is given by the inner product:

$$\widehat{f}(k) = \langle f(x), \phi_k(x) \rangle, \tag{3.23}$$

which is the Fourier coefficient at "frequency" k. Since the basis is complete, the function can be recovered by inverse transform:

$$f(x) = \sum_{k=0}^{\infty} \widehat{f}(k)\phi_k(x). \tag{3.24}$$

Rong et al. [Rong et al. 08] employed this transform to perform mesh editing on the base domain with low frequencies and reconstruct details with high frequencies. The eigenfunctions have been used for shape representation. In [Rustamov 07], a point signature, global point signature, was proposed, which consists of scaled Laplace-Beltrami eigenfunctions:

$$\text{GPS}(x) = (\frac{\phi_1(x)}{\sqrt{\lambda_1}}, \frac{\phi_2(x)}{\sqrt{\lambda_2}}, \ldots, \frac{\phi_k(x)}{\sqrt{\lambda_k}}, \ldots). \tag{3.25}$$

The inner product in this embedded signature domain corresponds to the Green's function of the Laplace-Beltrami operator

$$\langle \text{GPS}(x), \text{GPS}(y) \rangle = \sum_{k=1}^{\infty} \frac{\phi_k(x)\phi_k(y)}{\lambda_k}. \tag{3.26}$$

In [Ovsjanikov et al. 08], it was used to detect global intrinsic symmetries. The eigenfunctions were also applied to segmentation [Liu and Zhang 07] and registration [Reuter 10]. A set of Laplace-Beltrami eigenvalues $\{\lambda_k(x)\}$ is another signature of a shape. In [Reuter et al. 06], this set is named as "shape-DNA", which means two shapes can be compared through their Laplace-Beltrami eigenvalues.

3.3.3 Distances

Furthermore, quantities derived from the Laplace-Beltrami eigen-system lead to more technical tools and applications for shape analysis. One such quantity is the diffusion distance, defined as

$$d_t(x, y)^2 = \sum_{k=0}^{\infty} e^{-\lambda_k t}(\phi_k(x) - \phi_k(y))^2. \tag{3.27}$$

It is related to connectivity of two points by the number of their paths with a certain length [Coifman and Lafon 06], which represents the distance in the heat diffusion sense. Intuitively speaking, more paths between two points tend to produce a shorter diffusion distance. It has been used to define the diffusion map [Lafon et al. 06] that is an isometric embedding of a manifold. Bronstein et al. adopted diffusion distance as an isometry-invariant metric in shape recognition [Bronstein and Bronstein 11] and shape matching [Bronstein et al. 10b]. Different from the geodesic distance that is the shortest path, it considers all connected paths between the two points. Therefore, the diffusion distance characterizes shape geometry in a multiscale manner. In addition, it has also been applied to data fusion [Lafon et al. 06], gradient approximation [Luo et al. 09], and texture synthesis [Lu et al. 09]. Heat kernel is another widely-used quantity. The relation between diffusion distance and heat kernel can be expressed as

$$d_t(x, y)^2 = h_t(x, x) + h_t(y, y) - 2k_t(x, y). \tag{3.28}$$

Sun et al. [Sun et al. 09] proposed the heat kernel signature for finding multiscale features on manifold. The heat kernel signature measures the heat kernel from one point to itself at time t, which has shown to be flexible and stable in locating geometric features. In [Bronstein and Kokkinos 10], a scale-invariant heat kernel signature was proposed using Fourier transformation to eliminate the scale factor, and applied to shape retrieval. In [Ovsjanikov et al. 10], the heat kernel was used as a metric to define the heat kernel map for shape matching, which does not consider geometric compatibility of feature tuples. In [Jones et al. 08], the heat kernel was applied to local parameterization by heat triangulation. In [Hou and Qin 12b], heat kernel coordinates were introduced for dense registration of partial nonrigid shapes. They are local coordinates aligned by feature anchors.

In [Lipman et al. 10], the biharmonic distance was proposed for measuring distance on shapes. It is defined as

$$d_B(x, y)^2 = \sum_{k=1}^{\infty} \frac{(\phi_k(x) - \phi_k(y))^2}{\lambda_k^2}. \tag{3.29}$$

It has a similar formulation with the commute-time distance (resistance distance) [Fouss et al. 07]

$$d_C(x, y)^2 = \sum_{k=1}^{\infty} \frac{(\phi_k(x) - \phi_k(y))^2}{\lambda_k}. \tag{3.30}$$

The commute-time distance between two vertices on the graph is the average time it takes a random walk to go from one vertex to the other and back. The relation between the commute-time distance and the Green's function g_C is given by

$$d_C(x,y)^2 = g_C(x,x) + g_C(y,y) - 2g_C(x,y). \qquad (3.31)$$

Similarly, the relation between the biharmonic distance and the Green's function of the biharmonic operator can be expressed as

$$d_B(x,y)^2 = g_B(x,x) + g_B(y,y) - 2g_B(x,y), \qquad (3.32)$$

where $g_B(x,y)$ is the Green's function of the biharmonic operator Δ_M^2. It is somehow between geodesic distances for small scales and global shape-awareness for large scales. Applications of the biharmonic distance have been made to function interpolation and shape matching.

4

Admissible Diffusion Wavelets

As briefly introduced in Chapter 2, diffusion wavelets were developed in mathematics by [Coifman and Maggioni 06], which form a multiscale framework for data decomposition, compression, and analysis. This chapter presents an extended work of diffusion wavelets on 3D shapes. It aims to simplify the construction and advocate its use in computer graphics for space-frequency processing of shape geometry and value functions.

4.1   Diffusion Operator

Similar to the diffusion wavelets, the construction also follows a bottom-up manner, which starts from a local diffusion-type operator T and expands via its dyadic powers. For a meshed surface M, this operator forms a matrix with rows defined as functions at the associated vertices

$$\begin{cases} T(x,y) = A(y)e^{-\frac{\|x-y\|^2}{4t}}, & x \neq y \\ T(x,x) = \sum_{y \in M} T(x,y), & \text{otherwise} \end{cases} \tag{4.1}$$

where $A(y)$ is the vertex area of y, and t is a fixed quantity. It can be interpreted as a transition matrix of a random walk, which measures the probability of a one-step random walk moving from x to y on M. This explains that our wavelet operator has a diffusion-type distribution, which spreads from a vertex to its neighbors according to their distances. The quantity t corresponds to the neighborhood size. In [Hou and Qin 13], we set t as the half of the square of the average edge length. In this way, only a few neighboring entries $T(x,y)$ for vertex x have non-zero values, while other entries are all becoming zero. This indicates that T is a highly sparse matrix. For a uniformly sampled mesh, the operator T is approximately symmetric, since vertex areas are close.

This operator is closely related to the discrete Laplace operator L in [Belkin et al. 08], with an explicit expression

$$T = 4\pi t^2(L - 2D_L), \tag{4.2}$$

where D_L is the diagonal matrix with $D_L(x,x) = L(x,x)$.

Proof: Consider the definition of the mesh Laplace operator in [Belkin et al. 08]. Let K be a mesh in $\mathbb{R}^3$. Given a face, a mesh, or a surface X, let $Area(X)$ denote the area of X. For a face $t \in K$, the number of vertices in t is $\#t$, and $V(t)$ is the vertex set of t. The mesh Laplace operator L_K^h applied to a function $f : V \to \mathbb{R}$ is

$$L_K^h f(w) = \frac{1}{4\pi h^2} \sum_{t \in K} \frac{Area(t)}{\#t} \sum_{p \in V(t)} e^{-\frac{\|p-w\|^2}{4h}} (f(p) - f(w)), \qquad (4.3)$$

where h is a positive quantity.

We denote a set of triangles adjacent to p as $T(p)$. Now consider a point p in the computation of $L_K^h f(w)$, where all triangles in $T(p)$ are involved. The sum of these items is computed by

$$e^{-\frac{\|p-w\|^2}{4h}} (f(p) - f(w)) \sum_{t \in T(p)} \frac{Area(t)}{\#t}. \qquad (4.4)$$

It is easy to see that $\sum_{t \in T(p)} \frac{Area(t)}{\#t}$ is a formula to compute the vertex area of p, which is denoted as $A(p)$. Now Eq. (4.3) becomes

$$L_K^h f(w) = \frac{1}{4\pi h^2} \sum_{p \in V} A(p) e^{-\frac{\|p-w\|^2}{4h}} (f(p) - f(w)), \qquad (4.5)$$

where the Laplace operator L_k^h is

$$\begin{cases} L_K^h(w,p) = \frac{1}{4\pi h^2} A(p) e^{-\frac{\|p-w\|^2}{4h}}, & w \neq p \\ L_K^h(w,w) = -\sum_{p \in V} L_K^h(w,p), & \text{otherwise} \end{cases} \qquad (4.6)$$

It may be noted that we now can replace V, L_K^h, h, w, p with M, L, t, x, y, respectively:

$$\begin{cases} L(x,y) = \frac{1}{4\pi t^2} A(y) e^{-\frac{\|x-y\|^2}{4t}}, & x \neq y \\ L(x,x) = -\sum_{y \in M} L(x,y), & \text{otherwise} \end{cases} \qquad (4.7)$$

The operator T becomes

$$\begin{cases} T(x,y) = 4\pi t^2 L(x,y), & x \neq y \\ T(x,x) = -4\pi t^2 L(x,x), & \text{otherwise} \end{cases} \qquad (4.8)$$

Hence, we have

$$T = 4\pi t^2 (L - 2D_L), \qquad (4.9)$$

where D_L is the diagonal matrix with $D_L(x,x) = L(x,x)$. □

According to the algorithm for computing the Laplace operator on point clouds [Belkin et al. 09], the proposed operator T can be applied to point clouds too. In [Hou and Qin 13], we seek 10 nearest neighbors for each point, and only use them in non-zero entries of the operator at that point. The average edge length turns to the average of neighboring-point distances.

4.2 Wavelet Construction

4.2.1 Scaling Functions and Wavelets

In the diffusion wavelets, the dyadic powers of a diffusion operator are employed to smooth the function space $L^2(M)$. Decomposed from the dyadic powers, scaling functions form an orthogonal and complete basis in each subspace. The powers have decreasing ranks; therefore, the space is compressed. By dropping the orthogonality, the biorthogonal diffusion wavelets improve the scaling functions to be locally-supported. They still apply the rank-revealing QR decomposition to obtain a compressed space.

We adopt this bottom-up approach but with a different formulation when defining wavelets. Initially, the function space has a canonical basis $\Phi_0 = \{\varphi_{0,x}\}_{x \in M}$ of delta functions

$$\varphi_{0,x}(y) = \delta_x(y). \tag{4.10}$$

In the first level, the scaling functions $\Phi_1 = \{\varphi_{1,x}\}_{x \in M}$ are dilated once by T, given by

$$\Phi_1 = [\Phi_0 T]_r, \tag{4.11}$$

where $[\cdot]_r$ denotes the row-based normalization

$$\varphi_{1,x} = \frac{T(x, \cdot)}{\|T(x, \cdot)\|_1}. \tag{4.12}$$

Following this paradigm, the scaling functions in the j-th level

$$\Phi_j = [\Phi_{j-1} \Phi_{j-1}]_r$$

are constructed by the square of scaling functions in the $(j-1)$-th level

$$\varphi_{j,x} = [\varphi_{j-1,x} \Phi_{j-1}]_r. \tag{4.13}$$

They are dilated by $T^{2^{j-1}}$ subject to row-based normalization. The constructed scaling functions are locally-supported, non-negative

$$\varphi_{j,x}(y) \geq 0, \tag{4.14}$$

and normalized

$$\sum_{y \in M} \varphi_{j,x}(y) = 1. \tag{4.15}$$

The dyadic powers T^{2^j} decrease in rank as j increases, which indicates a compression of the function space. According to biorthogonal diffusion wavelets by [Maggioni et al. 05], Φ_j contains a Riesz basis with full rank spanning the subspace of level j, which is biorthogonal.

The wavelets are defined as differences of adjacent scaling functions:

$$\Psi_j = \Phi_{j-1} - \Phi_j,$$

with

$$\Psi_j = \Phi_{j-1} - \Phi_j, \quad j = 1, 2, \cdots \tag{4.16}$$

which are also locally-supported. This formulation is fundamentally different from the diffusion wavelets and the biorthogonal diffusion wavelets that both acquire the wavelets from the complement space $(I - \Phi\Phi^*)$ via QR decomposition. It is much simpler to compute. According to Eq. (4.15) and Eq. (4.16), the wavelet $\psi_{j,x}$ has a zero mean

$$\sum_{y \in M} \psi_{j,x}(y) = 0, \tag{4.17}$$

which implies that it vanishes at the zero frequency in its Fourier transform.

The index j in the above-formulated functions connects with the "frequency" in spectral domain. Small values of j correspond to high frequencies, while large values correspond to low frequencies. As j increases, the scaling functions and wavelet functions dilate to larger areas rapidly because of the dyadic powers. For large enough values of j, scaling functions converge to a constant

$$\lim_{j \to \infty} \varphi_{j,x} = \frac{1}{A(M)}, \tag{4.18}$$

where $A(M)$ denotes the total surface area of M. Accordingly, wavelets converge to zero

$$\lim_{j \to \infty} \psi_{j,x} = 0. \tag{4.19}$$

Fig. 4.1 illustrates some scaling functions and wavelets on a 1D manifold with 100 points. The modified diffusion wavelets have nice shapes of scaling functions and wavelets. The wavelets are oscillating and attenuated over the domain.

We have shown that the admissible diffusion wavelets can be computed on both meshes and point clouds, which only involves local geometry. Therefore, it can be directly generalized to handle other data domains such as graphs with discrete structure, tensor fields defined over multi-dimensional volume, and even higher-dimensional scientific data with curved manifold structure.

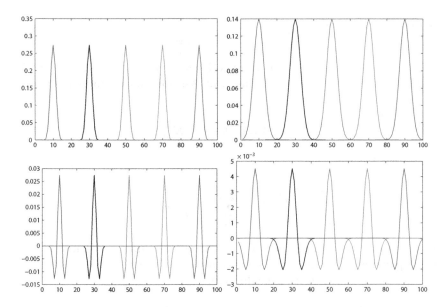

Figure 4.1. Examples of admissible diffusion wavelets: scaling functions (top) and wavelets (bottom).

4.2.2 Admissibility Condition

Next, we will have a brief discussion on the admissibility condition. This condition is for continuous wavelet transform originally in Euclidean space. Here, we simply extend it to manifold space by the following definition. On manifold with bounded geometry, a wavelet ψ is admissible, or equivalently satisfies the admissibility condition, if

$$\sum_{k=0}^{N} \frac{|\widehat{\psi}(k)|^2}{k} < \infty,$$

where $\widehat{\psi}(k)$ denotes the manifold harmonic transform [Vallet and Lévy 08] of wavelet ψ, and N is the data size. This definition is analogous to its origin in Euclidean space, since the manifold harmonic transform is a manifold Fourier transform. The wavelets $\psi_{j,x}$ defined by Eq. (4.16) are admissible.

Proof: Recall that the zero-frequency manifold harmonic basis

$$\phi_0 = \frac{1}{\sqrt{A(M)}},$$

which is a constant everywhere on M. According to Eq. (4.17), the wavelet $\psi_{j,x}$ has zero mean. Hence, it vanishes at zero frequency in its manifold harmonic transform,

$$\widehat{\psi}_{j,x}(0) = \langle \psi_{j,x}, \phi_0 \rangle = 0.$$

Also, since the wavelet $\psi_{j,x}$ is compactly-supported, it has limited band-width in its manifold harmonic transform. Assume that its upper frequency is $K(j) < N$. We then have

$$\widehat{\psi}_{j,x}(k) = 0, \text{ for } k > K(j).$$

Hence, we have

$$\sum_{k=0}^{\infty} \frac{|\widehat{\psi}_{j,x}(k)|^2}{k} = \sum_{k=1}^{K(j)} \frac{|\widehat{\psi}_{j,x}(k)|^2}{k} < \infty.$$

In the worst case, this sum is cut off at the maximal frequency $N-1$ corresponding to the maximal eigenvalue λ_{N-1}, which is still bounded to a finite number. Hence, $\psi_{j,x}$ is admissible.   □

In fact, by recalling [Antoine and Vandergheynst 99], the zero-mean property of $\psi_{j,x}$ suffices the admissibility condition for compactly-supported wavelets. This condition is critical for wavelet transform, as it ensures the transform can be fully recovered.

Fig. 4.2 visualizes a scaling function and a wavelet of the admissible diffusion wavelets on a meshed surface by color coding. Scaling functions and wavelets are oscillating and attenuated on the shape. As the scale increases, the scaling functions dilate to larger areas.

4.3   Wavelet Transform

Wavelet transform and inverse-transform (or reconstruction) are used to analyze and edit functions in both space and frequency domains.

4.3.1   Transform

The scaling and wavelet transforms (denoted by $\mathcal{S}_f$ and $\mathcal{W}_f$, respectively) of a function $f(x) \in L^2(M)$ with $x \in M$ are computed by the inner product over the domain M:

$$\begin{cases} \mathcal{S}_f(j,x) = \langle \varphi_{j,x}, f \rangle \\ \mathcal{W}_f(j,x) = \langle \psi_{j,x}, f \rangle \end{cases} \tag{4.20}$$

where j and x localize the frequency domain and the space domain, re-spectively. The scaling coefficient $\mathcal{S}_f(j,x)$ is a smoothed representation of

Figure 4.2. Rendering a scaling function and a wavelet of the admissible diffusion wavelets on a meshed surface by color coding. See color insert.

function f, which is an approximation to f at scale j. The wavelet coefficient $W_f(j, x)$ records the residual detailed information of f with respect to the scale j.

Fig. 4.3 shows mutiresolution decomposition [Mallat 89], by illustrating some scaling and wavelet coefficients of a 1D function. The function is smoothed by the scaling transform, with details recorded in the wavelet transform at different scales. In Fig. 4.4, a shape and a scalar field (mean curvature map) on the shape are transformed to different scales by our scaling transform. The scaling coefficients become constant everywhere at the coarsest scale. This indicates that the function space is highly compressible after scaling transforms, as it is gradually smoothed. In [Hou and Qin 13], we retained full resolution of the scaling and wavelet transform, with our emphasis on accurate high-frequency processing. For applications with other purposes such as level-of-details rendering and function approximation, one could compress the scaling coefficients (i.e., smoothed functions) via down-sampling.

4.3.2 Reconstruction

The reconstruction (inverse transform) aims to recover a function from its coefficients. For the diffusion wavelets, the reconstruction uses the same basis as the transform, since it is orthogonal and complete. The scaling functions of the biorthgonal diffusion wavelets form a Riesz basis. According to [Christensen 02], if $\{\varphi_i\}$ is a Riesz basis, there is a unique dual basis $\{\widetilde{\varphi}_j\}$ that is orthogonal to $\{\varphi_i\}$:

$$\langle \varphi_i, \widetilde{\varphi}_j \rangle = \delta_{ij}, \tag{4.21}$$

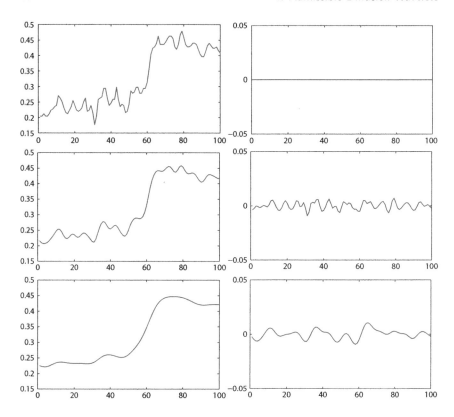

Figure 4.3. Scaling (left) and wavelet (right) coefficients of a 1D function at $j = 0, 3, 6$. The function is smoothed by the scaling transform, with details recorded in the wavelet transform at different scales.

and $\{\varphi_i\}$ is biorthogonal. The reconstruction is given by [Maggioni et al. 05]

$$f = \sum_{x \in M} \langle f, \varphi_{j,x} \rangle \widetilde{\varphi}_{j,x}, \qquad (4.22)$$

where $\{\widetilde{\varphi}_{j,x}\}$ is the dual basis of $\{\varphi_{j,x}\}$. This requires the scaling matrix Φ_j to have full rank. Therefore, the rank-revealing QR decomposition is necessary. The dual basis is computed by the matrix inverse

$$\Phi_j \widetilde{\Phi}_j = I. \qquad (4.23)$$

Different from the diffusion wavelets and the biorthgonal diffusion wavelets, we adopt a rapid reconstruction by utilizing the fact that the redundant scaling functions and wavelets are all with full-resolution. It is a superposition of wavelet and scaling coefficients. According to the definition

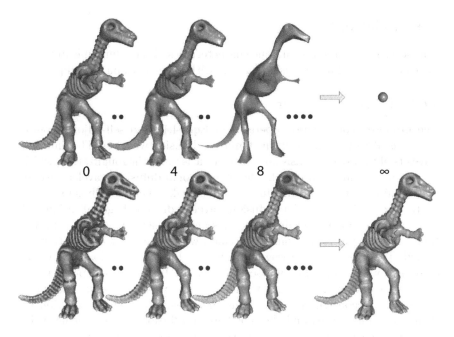

Figure 4.4. Multiscale representations of a shape (top) and a scalar field defined on the shape (bottom) by our scaling transform. The scaling coefficients become constant everywhere at the coarsest scale.

of our wavelets, we have

$$
\begin{aligned}
\mathcal{W}_f(j, x) &= \langle f, \psi_{j,x} \rangle \\
&= \langle f, \varphi_{j-1,x} - \varphi_{j,x} \rangle \\
&= \mathcal{S}_f(j-1, x) - \mathcal{S}_f(j, x),
\end{aligned}
$$

for $j = 1, 2, \dots$. This implies that the wavelet coefficients are full-resolution details at different levels of frequencies. The function can be rapidly reconstructed by a series of J levels of wavelet coefficients $\{\mathcal{W}_f(j)\}_{j=1}^{J}$ and a scaling coefficient at the coarse level $\mathcal{S}_f(J)$,

$$
f(x) = \mathcal{S}_f(0, x) = \mathcal{S}_f(J, x) + \sum_{j=1}^{J} \mathcal{W}_f(j, x). \tag{4.24}
$$

Since all the coefficients have full resolution, the rapid reconstruction is lossless.

4.4   Relations

This section highlights the unique characteristics of the admissible diffusion wavelets by revealing intrinsic relations with other relevant techniques.

4.4.1   Diffusion Wavelets

The proposed construction operator in Eq. (4.1) is a self-adjoint operator. Thus, it can be placed in the family of diffusion wavelets. The diffusion wavelets obtain scaling basis and wavelet basis via orthogonalization, which are not well locally-supported. The biorthogonal diffusion wavelets release this requirement for scaling functions to make them locally-supported. Furthermore, the admissible diffusion wavelets drop orthogonalization and downsampling for both scaling functions and wavelets. The obtained functions are with full resolution at all scales, with redundant information included. This costs extra storage but gives us fast construction of wavelets at some low levels in the multiscale hierarchy. In the second part of this book, we will show that a small number of low level scales suffice the applications. For reconstruction, the diffusion wavelets recover the function by their orthogonal basis, and the biorthgonal diffusion wavelets achieve the same goal by the biorthogonal dual basis. The admissible diffusion wavelets adopt a rapid reconstruction that does not rely on any basis. The differences in wavelet design result from different application purposes. Diffusion wavelets are often used for data compression and function approximation, while the admissible diffusion wavelets are used for space-frequency analysis of 3D shapes at some small scales.

4.4.2   Manifold Harmonic Basis

Analogous to sine/cosine functions in Fourier transform, manifold harmonics (Laplace-Beltrami eigenfunctions) are global functions on manifolds, which support spectral analysis of functions in L^2. The construction of manifold harmonic basis follows a top-down approach that starts from large scales to small scales. Low-frequency basis functions are fast to compute, while high-frequency basis functions are time consuming. On the contrary, the construction of the admissible diffusion wavelets adopts a bottom-up approach, where high-frequency basis functions are fast to compute. From the perspective of running time, the admissible diffusion wavelets are more efficient for high-frequency processing in spectral analysis. This feature is explained in Fig. 4.5. The reconstruction of manifold harmonic transform is lossless only if the entire spectrum of all eigenfunctions is utilized, which is extremely expensive to compute. The common practice for manifold harmonic transform is that high-frequency basis functions are oftentimes ignored, leading to lossy reconstruction and information processing.

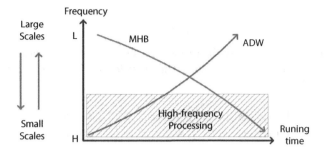

Figure 4.5. The construction of the manifold harmonic basis (MHB) uses a top-down approach starting from large scales to small scales. Low-frequency basis is fast to compute, while high-frequency basis is time-consuming. On the contrary, the construction of the admissible diffusion wavelets (ADW) adopts a bottom-up approach, where high-frequency basis is fast to compute. Therefore, the admissible diffusion wavelets are more efficient for high-frequency processing in shape analysis.

The manifold harmonic transform is easy to manipulate shape signals in low frequencies, while operations in high frequencies always consume extremely high computational cost. In sharp contrast, the reconstruction of the admissible diffusion wavelets has an "inverse" procedure that starts from high frequencies, leaving low frequency components in the scaling coefficients at the coarsest level. This bottom-up approach makes it extremely attractive and powerful when processing shape signals in fine, local details.

4.4.3   Spectral Graph Wavelets

As introduced in Chapter 2, the spectral graph wavelets are constructed using the graph Fourier transform. In graph Fourier domain, the wavelets have explicit formulations generated by a kernel and dilated by time t. It leads to a generic mechanism for generating continuous wavelets. If one chooses a diffusion kernel (e.g. the heat kernel) to generate the wavelets, they will be similar with the admissible diffusion wavelets in function shape and application effects. Same as the manifold harmonics basis, the computation of spectral graph wavelets needs to solve the eigen-decomposition of the graph Laplace operator.

4.4.4   Differential Coordinates

Differential coordinates [Alexa 03] have been extensively studied and applied to fast mesh editing. The Laplacian operator is a linear operator that captures details of a surface. Given the divergence of the vector field b, the scalar field f can be reconstructed by solving the Poisson equation

[Yu et al. 04]

$$\Delta f = b,$$

which gives rise to a sparse linear system. Our wavelet transform $\mathcal{W}_f(1, x)$ has very similar behaviors, since the wavelet basis Ψ_1 is a differential operator with entries

$$\begin{cases} \Psi_1(x, y) = -\frac{T(x,y)}{2T(x,x)}, & x \neq y \\ \Psi_1(x, x) = -\sum_x \Psi_1(x, y) = \frac{1}{2}, & \text{otherwise} \end{cases} \qquad (4.25)$$

where $T(x, x) = \sum_y T(x, y)$ as defined in Eq. (4.1). When the field f is a function of coordinates, $\mathcal{W}_f(1, x)$ is a representation of differential coordinates. For wavelet transform at high frequencies, they behave exactly like differential coordinates in larger neighborhood [Lipman et al. 04].

4.5   Space-Frequency Processing Framework

We have detailed the admissible diffusion wavelets as powerful and effective tools for space-frequency processing with a suite of applications in visual computing. The admissible diffusion wavelets are constructed in a bottom-up manner, which starts from a local operator T and expands as its dyadic powers increase. We have formulated the scaling functions and wavelets (that are also locally-supported) in a mathematically rigorous way. The admissibility condition is naturally enforced by the definition of wavelets. The rapid reconstruction is carried out by employing coefficients, located at multiple frequencies. The dyadic power series T^{2^j} efficiently scales the Hilbert space. As j increases, the admissible diffusion wavelets collect valuable information from small scales, propagate that to large scales, and continue to spread through the hierarchy. The rapid computation of sparse admissible diffusion wavelets at the first several levels of scales makes them very efficient for high-frequency processing. For processing in low frequencies (i.e., large scales), the computational efficiency might become moderate. Therefore, it makes sense to concentrate on high-frequency-relevant applications such as saliency visualization, feature definition and extraction, and geometry analysis and processing. In practice, since the dyadic power grows very fast, several levels of admissible diffusion wavelets appear to be adequate for our applications, where the computation is significantly faster than solving the global eigen-system. More importantly, spatial localization of wavelets empowers the admissible diffusion wavelets to have the unique characteristic of space-frequency processing towards seeking local features and filtering local geometry, while retaining gross shape globally.

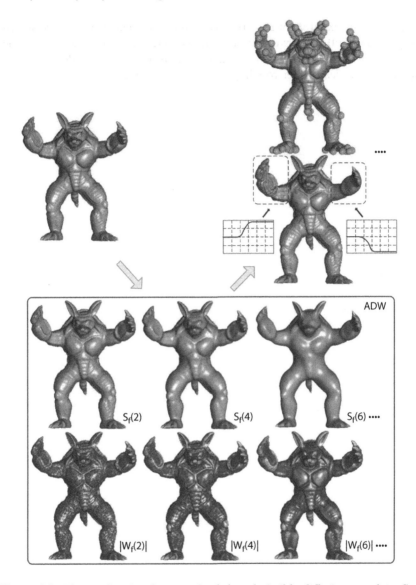

Figure 4.6. The application framework of the admissible diffusion wavelets. See color insert.

Fig. 4.6 illustrates the framework of space-frequency processing by the admissible diffusion wavelets. A shape is transformed to scaling coefficients $\mathcal{S}_f$ visualized as smoothed shapes and wavelet coefficients $\mathcal{W}_f$ visualized as coded colors. Examples are shown to portray the application-relevant results. In the example of feature extraction, multiscale features are shown as green balls. In the example of local geometry filtering, one arm of the armadillo is smoothed while the other is enhanced. For more details of the applications, please refer to Chapter 10 and Chapter 11.

5

Mexican Hat Wavelet

This chapter presents the Mexican hat wavelet defined on manifold data. This continuous wavelet is rigorously derived from the heat kernel by taking the negative first-order derivative with respect to time. As a solution to the heat equation, it has a clear initial condition: the Laplace-Beltrami operator. Following a popular methodology in mathematics, we analyze the wavelet and its transforms from a Fourier perspective. By formulating Fourier transforms of bivariate kernels and convolutions, we obtain its explicit expression in the Fourier domain, which is a scaled differential operator continuously dilated via heat diffusion. The wavelet has localization in both space and frequency, which enables space-frequency analysis of input functions.

5.1 Manifold Harmonics

One long-lasting task in geometry processing is to develop functional analysis tools on curved surfaces. Without Euclidean metric, it is extremely challenging to explicitly define functions on manifolds. Many existing methods are hinged upon differential geometry, where surface parameterization is frequently unavoidable. In [Antoine et al. 10, Hou and Qin 12a], wavelets were designed in frequency domain via spectral decomposition. Functions that have no closed-form expression on manifolds may have analytical formulations in frequency domain.

Fourier transform is well known for spectral analysis. Local areas of curved surfaces are homogeneous to 2D planar patches, where the Euclidean Fourier transform can be applied for spectral processing [Pauly and Gross 01]. In terms of adapting the Fourier transform on manifolds, basis functions are critical for orthogonally decomposing the space to a series of shape spectra. In [Ben-Chen and Gotsman 05, Karni and Gotsman 00], eigenfunctions of the symmetric Laplacian of the connectivity graph are adopted as a Fourier basis, which is derived from the mesh topology but not the geometry. Analogous to the Fourier basis in Euclidean metric, manifolds have similar orthonormal basis formed by eigenfunctions of the Laplace-Beltrami operator [Lévy 06]. Accordingly, Vallet and Lévy [Vallet and Lévy 08] defined the manifold harmonic transform that is a fully adapted manifold Fourier transform, expanded on manifold harmonics (i.e., Laplace-Beltrami eigenfunctions). For applications, Rong et al. [Rong et al. 08] employed this spectral decomposition to perform mesh editing on the base domain with low frequencies and reconstruct details with high frequencies. The Fourier basis, consisting of functions repeatedly oscillating over the entire domain, does not have localization in space. Therefore, adapted Fourier transforms only allow global operations of input functions.

In Euclidean domain, the Fourier transform of an integrable function $f(x)$ is defined as

$$\widehat{f}(\omega) = \int_{-\infty}^{\infty} f(x)e^{-i\omega x}dx, \tag{5.1}$$

where $e^{-i\omega x}$ is the Fourier basis, and ω is the frequency. By Euler's formula, the Fourier basis ends up with sine and cosine functions, which are orthonormal and oscillating over the domain. They can be interpreted as eigenfunctions of derivative operators with respect to x. For example, $e^{-i\omega x}$ is the eigenfunction of the second-order derivative operator associated with the eigenvalue ω^2, because of

$$\frac{d^2}{dx^2}e^{-i\omega x} = \omega^2 e^{-i\omega x}. \tag{5.2}$$

That is, by taking the Fourier basis as eigenfunctions of the second-order derivative operator, the eigenvalues are real numbers corresponding to the squares of frequencies.

For a compact manifold M, the Laplace-Beltrami operator Δ_M is negative and formally self-adjoint, which has a well-defined eigen-system $\{\lambda_k, \phi_k\}$:

$$\Delta_M \phi_k(x) = -\lambda_k \phi_k(x), \tag{5.3}$$

where λ_k and $\phi_k(x)$ are the k-th eigenvalue and the k-th eigenfunction, respectively. Analogous to the Fourier basis in Euclidean domain, eigenfunctions $\{\phi_k\}$ form an orthogonal and complete basis for the Hilbert space $L^2(M)$, which are also called manifold harmonics. A function $f(x) \in L^2(M)$ can be expanded on this basis,

$$f(x) = \sum_{k=0}^{\infty} \widehat{f}(k)\phi_k(x), \tag{5.4}$$

where

$$\widehat{f}(k) = \langle f(x), \phi_k(x) \rangle. \tag{5.5}$$

Eq.(5.4) is the spectral decomposition of function $f(x)$, and Eq.(5.5) is the manifold harmonic transform [Vallet and Lévy 08]. With a slight abuse of language, we call $\widehat{f}(k)$ the Fourier transform, or Fourier coefficient, of function $f(x)$ on manifold M. Consequently, Eq. (5.4) is called the inverse Fourier transform on manifolds.

5.2 Bivariate Kernels and Convolutions

In this section, we introduce Fourier transforms of bivariate kernels and convolutions on manifolds, which are necessary in wavelet design.

With the purpose for space-frequency analysis and functional analysis for novel wavelet formulation, we consider a bivariate kernel $\theta(x, y)$ of a self-adjoint operator Θ, and explicitly employ such kernel to compute the convolution

$$\theta(x, y) * f(x) = \Theta f(x) = \langle \theta(x, y), f(y) \rangle. \tag{5.6}$$

The bivariate kernel can be expanded on the Fourier basis

$$\theta(x, y) = \sum_{k=0}^{\infty} \widehat{\theta}(k) \phi_k(x) \phi_k(y), \tag{5.7}$$

where $\widehat{\theta}(k)$ is the Fourier transform of kernel $\theta(x, y)$ given by

$$\widehat{\theta}(k) = \langle \langle \theta(x, y), \phi_k(x) \rangle, \phi_k(y) \rangle. \tag{5.8}$$

Note that off-diagonal entries in frequency are ignored, because of the orthogonality of the Fourier basis.

Table 5.1 documents some important examples of functions and kernels under the Fourier transform. A Dirac function and a basis function are

Function / Kernel	Space	Fourier
Dirac function	$\delta_x(y)$	$\phi_k(x)$
Fourier basis function	$\phi_j(x)$	$\delta_j(k)$
Convolution	$\theta(x, y) * f(x)$	$\widehat{\theta}(k) \widehat{f}(k)$
Laplace-Beltrami kernel	$\Delta_M(x, y)$	λ_k
Heat kernel	$h_t(x, y)$	$e^{-\lambda_k t}$
Mexican hat wavelet	$\psi_t(x, y)$	$\lambda_k e^{-\lambda_k t}$
Biharmonic wavelet kernel	$\psi_t^2(x, y)$	$\lambda_k^2 e^{-\lambda_k t}$

Table 5.1. Some manifold functions and kernels under the Fourier transform.

dualities under the Fourier transform. The impulse function in space has infinite bands in frequency, and vice versa.

We extend the Fourier transform to the convolution by the following theorem.

Theorem 5.1 (Convolution Theorem) *The Fourier transform of a convolution $\theta(x, y) * f(x)$ is the point-wise product of Fourier transforms $\widehat{\theta}(k)\widehat{f}(k)$.*

Proof: By definition, we have the Fourier transform of the convolution

$$
\begin{aligned}
\langle\langle\theta(x, y), f(y)\rangle, \phi_k(x)\rangle &= \langle\langle\theta(x, y), \phi_k(x)\rangle, f(y)\rangle \\
&= \langle\widehat{\theta}(k)\phi_k(y), f(y)\rangle \\
&= \widehat{\theta}(k)\widehat{f}(k).
\end{aligned} \tag{5.9}
$$

Hence, the statement is proved.   □

This theorem facilitates designing functions on manifolds in Fourier domain by concatenation. Functions with complex forms on manifolds may have simple expression in the Fourier domain. One can design new functions by concatenations of known functions in the Fourier domain. For example, our Mexican hat wavelet is formulated as the product of the Laplace-Beltrami kernel and the heat kernel in the Fourier domain:

$$
\widehat{\psi}_t(k) = \lambda_k e^{-\lambda_k t}.
$$

As introduced in [Hammond et al. 11], the graph Fourier transform satisfies the Parseval's relation. It is easy to see that this relation also holds on manifolds. We give the theorem and proof of this relation below for completeness.

Theorem 5.2 (Parseval's Theorem [Hammond et al. 11]) *For any two functions $f(x), g(x) \in L^2(M)$, the following relation holds:*

$$
\langle f(x), g(x)\rangle = \sum_{k=0}^{\infty} \widehat{f}(k)\widehat{g}(k). \tag{5.10}
$$

Proof: By definition, we have

$$
\begin{aligned}
\sum_{k=0}^{\infty} \widehat{f}(k)\widehat{g}(k) &= \sum_{k=0}^{\infty} \langle f(x), \phi_k(x)\rangle \widehat{g}(k) \\
&= \langle f(x), \sum_{k=0}^{\infty} \widehat{g}(k)\phi_k(x)\rangle \\
&= \langle f(x), g(x)\rangle.
\end{aligned} \tag{5.11}
$$

Hence, the statement is proved.   □

This theorem implies the energy conservation in space and frequency. It is an important property of the manifold Fourier transform, which will be used for the perturbation analysis of quantities defined in spectral domain.

5.3 Mexican Hat Wavelet

We advocate the Mexican hat wavelet on manifold geometry, which is rigorously derived from heat diffusion. As we know, the Mexican hat wavelet is defined as the negative normalized second derivative of a Gaussian $G_t(x)$ with respect to x. It is also defined as the negative first derivative of the Gaussian with respect to t. It is easy to understand, since the Gaussian is the fundamental solution to the heat equation

$$\frac{\partial u}{\partial t} = \nabla^2 u. \tag{5.12}$$

On manifolds, the heat equation is given by

$$\frac{\partial u(x,t)}{\partial t} - \Delta_M u(x,t) = 0, \tag{5.13}$$

where Δ_M is the Laplace-Beltrami operator. The fundamental solution to this PDE is known to be the heat kernel

$$h_t(x,y) : \mathbb{R}^+ \times M \times M \to \mathbb{R}^+,$$

given by formulations in space

$$h_t(x,y) = \sum_{k=0}^{\infty} e^{-\lambda_k t} \phi_k(x)\phi_k(y), \tag{5.14}$$

and Fourier domain

$$\widehat{h}_t(k) = e^{-\lambda_k t}. \tag{5.15}$$

The heat kernel in Fourier domain $\widehat{h}_t(k)$ is a Gaussian of $\sqrt{\lambda_k}$. This implies that, although the heat kernel has no closed-form expression in space, it has an explicit expression as a Gaussian in the Fourier domain.

Similarly, the Mexican hat wavelet

$$\psi_t(x,y) : \mathbb{R}^+ \times M \times M \to \mathbb{R}$$

on manifolds is defined as the negative first-order derivative of the heat kernel with respect to t

$$\psi_t(x,y) = -\frac{\partial h_t(x,y)}{\partial t} = \sum_{k=0}^{\infty} \lambda_k e^{-\lambda_k t} \phi_k(x)\phi_k(y). \tag{5.16}$$

Its Fourier transform is given by

$$\widehat{\psi}_t(k) = \lambda_k e^{-\lambda_k t} = \widehat{\Delta}_M(k)\widehat{h}_t(k). \tag{5.17}$$

It is a solution to the heat equation with the Laplace-Beltrami operator as the initial condition.

By defining Fourier transforms of bivariate kernels and convolutions, we further reveal that in the Fourier domain, the Mexican hat wavelet is a product of the Laplace-Beltrami operator and the heat kernel. It is, therefore, a scaled differential operator continuously dilated through heat diffusion. According to the Convolution Theorem 5.1, its Fourier transform is the product of the Laplace-Beltrami kernel and the heat kernel, as shown in Table 5.1. This means that it is a scaled differential operator dilated by heat diffusion.

Here, t is related to "frequency". Different from the frequency index k in Fourier transform, the index t is continuous, leading to a continuous wavelet. Small values of t correspond to high frequencies, while large values correspond to low frequencies, which is opposite of Fourier index k. In the same spirit of diffusion wavelets [Coifman and Maggioni 06], the Mexican hat wavelet uses diffusion for dilation and scaling. Diffusion wavelets smooth the space by discrete powers of a diffusion operator, while the wavelet performs this task by continuous heat diffusion subject to the heat equation. Driven by different applications, the diffusion wavelets do not have oscillating and attenuated shapes.

Fig. 5.1 shows two wavelet functions on a 1D manifold with 400 uniformly sampled points, and their Fourier transforms in frequency domain. The Mexican hat wavelet has Gaussian decays in both space and frequency.

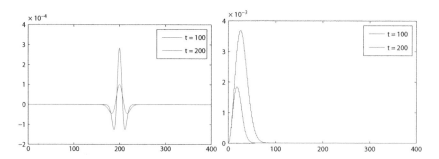

Figure 5.1. Two wavelet functions on a 1D manifold with 400 uniformly sampled points (left), and their Fourier transforms in frequency domain (right). For a larger scale, the Mexican hat wavelet has a wider window in space, but a narrower window in frequency.

Figure 5.2. Mexican hat wavelets on a box at $t = 10, 20, 40, 80, 160, 320$. The centers of functions are marked by black balls. The Mexican hat wavelet behaves differently at a corner point and a planar point. See color insert.

The figure illustrates that for a larger scale (red curves), the wavelet has a wider window in space, but a narrower window in frequency. This indicates both space and frequency resolutions cannot be arbitrarily high. The sampling of the wavelet in space and frequency follows the Heisenberg principle [Jaffard et al. 01]. Fig. 5.2 shows some Mexican hat wavelets on a box. The Mexican hat wavelet behaves differently at a corner point and a planar point. In Fig. 5.3, we show some wavelets on a 2D manifold by color-coding. The center points are shown as dark balls. The red-to-blue color illustrates the shown wavelets have similar shapes of the 1D wavelets, which are oscillating and attenuated on the manifold. Fig. 5.4 visualizes $\psi_t(x, x)$ at different values of t on deformable shapes. The MHW is relevant to shape geometry, and invariant to isometric deformation.

This above scheme for defining wavelets can be extended to other self-adjoint operators. For instance, we define the p-th Mexican hat wavelet as,

$$\psi_t^p(x, y) = \sum_{k=0}^{\infty} \lambda_k^p e^{-\lambda_k t} \phi_k(x) \phi_k(y). \tag{5.18}$$

The order $p > 0$ has a damping effect to the exponential attenuation of the wavelets in Fourier domain. Specifically, for $p = 2$, it is the biharmonic wavelet, with respect to the biharmonic operator $(\Delta)^2$, as shown in Table 5.1.

We have detailed the Mexican hat wavelet and its transforms on manifold geometry. It is solely derived from the heat equation. By means of formulating bivariate kernels for differential operators and their Fourier transforms, we obtain an explicit spectral expression of the wavelet. This strongly advocates a novel approach to developing functional analysis tools in Fourier domain. The wavelet has localization in both space and frequency, hence having a strong appeal to applications in space-frequency

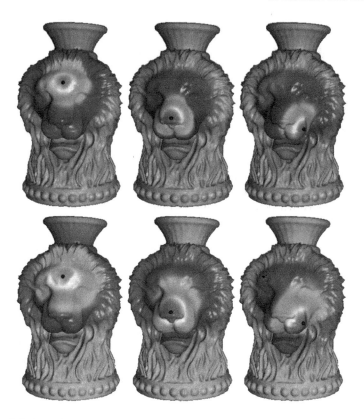

Figure 5.3. Some Mexican hat wavelets on a 2D manifold visualized by color plots, at $t = 50$ (top) and $t = 100$ (bottom). Wavelets are attenuated and oscillating on the manifold. See color insert.

analysis. The proposed Fourier method for computing convolutions of wavelet transforms is stable and efficient, which is consistent with the case in signal processing. Closely related to the Laplace-Beltrami operator, the wavelet captures geometric information up to isometry, which facilitiates more applications in shape analysis.

Several other issues that require further comprehensive studies include boundary conditions, eigenvalue problems for other types of initial-value partial differential equations and their corresponding operators, and more and broader applications. For partial shapes with boundaries, imposing stochastic completeness leads to inconsistent values of the wavelet near boundaries. And the affected area becomes larger when t increases. Fast solving the eigenvalue problem of the Laplace-Beltrami operator is another challenge. Particularly, for the Fourier transform and derived spectral expressions, they require the eigenfunctions to be strictly orthonormal.

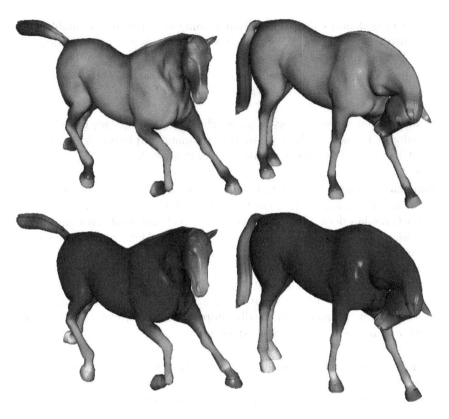

Figure 5.4. Visualizing $\psi_t(x, x)$ at $t = 80$ (Top) and $t = 640$ (Bottom) on deformable shapes. It is relevant to shape geometry, and invariant to isometric deformation. See color insert.

5.4   Properties

The Mexican hat wavelet has Gaussian decays in both space and frequency, which implies it can extract information in a space-frequency window. We further elaborate some important properties, such as admissibility, convergence, informativeness, and stability.

Symmetry   The Mexican hat wavelet is symmetric in space, i.e.,

$$\psi_t(x, y) = \psi_t(y, x). \tag{5.19}$$

Zero-Mean   The Mexican hat wavelet has a zero mean, given by

$$\int_M \psi_t(x, y) d\mu(y) = 0, \text{ for all } t > 0, \tag{5.20}$$

where $\mu(y)$ denotes the Riemannian volume of y on the manifold. This is a direct consequence of a property of the heat kernel on stochastically complete manifolds [Hsu 89],

$$\int_M h_t(x,y)d\mu(y) = 1, \text{ for all } t > 0. \tag{5.21}$$

It implies that the wavelet $\psi_t(x,y)$ vanishes at zero-frequency in its Fourier transform. In fact, as $\lambda_0 = 0$ for the Neumann Laplace-Beltrami operator [Chavel 84], the Fourier transform has $\widehat{\psi_t}(0) = 0$.

Gaussian-Decay   The Mexican hat wavelet has Gaussian decays in both space and frequency. In frequency, the Fourier transform $\widehat{\psi_t}(k) = \lambda_k e^{-\lambda_k t}$ has a Gaussian decay. In space, the heat kernel and its derivatives have a Gaussian upper bound [Boggess and Raich 10], which implies a Gaussian decay on the manifold with given t. Hence, the wavelet drops exponentially to zero along the manifold. This is also related to the multiscale property of the heat kernel, which states that for small value of t, the heat kernel $h_t(x, \cdot)$ is mostly determined by a small neighborhood of x. The Gaussian decay indicates that, although not locally-supported, the wavelet is localized in space and frequency [Roşca and Antoine 09].

Admissibility   The admissibility condition of wavelets is to ensure the function can be inversely recovered after transforms. We first formally define the admissibility condition for wavelets on manifolds as follows.

Definition 5.3 (Admissibility Condition) On compact Riemannian manifolds, a wavelet ψ is admissible, or equivalently satisfies the admissibility condition, if and only if

$$\sum_{k=0}^{\infty} \frac{|\widehat{\psi}(k)|^2}{k} < \infty.$$

Theorem 5.4 (Admissibility Theorem) *The Mexican hat wavelet defined in Eq. (5.16) is admissible.*

Proof: We have the Fourier transform of the Mexican hat wavelet:

$$\widehat{\psi_t}(k) = \lambda_k e^{-\lambda_k t}.$$

By recalling the Weyl's Theorem [Chavel 84], eigenvalues have an asymptotic formula. It yields that, for a 2-manifold M, the eigenvalue λ_k can be approximated by ck, with some positive constant c determined by $\mu(M)$. Therefore, we have

$$\frac{|\widehat{\psi}(k)|^2}{k} = \frac{\lambda_k^2 e^{-2\lambda_k t}}{k} < c_1 k e^{-c_2 k t},$$

with positive constants c_1 and c_2, and

$$\sum_{k=0}^{\infty} \frac{|\widehat{\psi}(k)|^2}{k} < \sum_{k=0}^{\infty} c_1 k e^{-c_2 k t} < \frac{c_1}{c_2 t} < \infty.$$

Hence, the Mexican hat wavelet is admissible. □

Convergence The long-time $\psi_t(x, y)$ converges to a stable state,

$$\lim_{t \to \infty} \psi_t(x, y) = 0, \text{ for all } x, y \in M. \tag{5.22}$$

It means for large value of t, $\psi_t(x, y)$ converges to zero everywhere on the manifold, which is the state of zero-frequency. It is based on the fact that, for a manifold with bounded geometry $\mu(M) < \infty$, we have a stable state of the heat kernel [Grigor'yan 06]

$$\lim_{t \to \infty} h_t(x, y) = \frac{1}{\mu(M)}, \text{ for all } x, y \in M. \tag{5.23}$$

which is constant everywhere on M.

The short-time $\psi_t(x, y)$ converges to the Laplace-Beltrami kernel

$$\lim_{t \to 0} \psi_t(x, y) = \sum_{k=0}^{\infty} \lambda_k \phi_k(x) \phi_k(y) = \Delta_M(x, y). \tag{5.24}$$

This convergence once again demonstrates the deep bond between the Mexican hat wavelet and the Laplace-Beltrami operator.

Informativeness Similar to the heat kernel, the wavelet $\psi_t(x, y)$ is also informative. This property is abstracted as the following statement.

Theorem 5.5 (Informative Theorem) *Let $T : M \to N$ be a bijective map between manifolds M and N. If*

$$\psi_t^M(x, y) = \psi_t^N(T(x), T(y))$$

for all $x, y \in M$ and all $t > 0$, then we have

$$h_t^M(x, y) = h_t^N(T(x), T(y))$$

and T is isometric.

Proof: By definition, we have

$$\int_0^t \psi_t(x, y) dt = h_0(x, y) - h_t(x, y).$$

By the Dirac condition [Grigor'yan 06], we have

$$h_0(x, y) = \delta(x, y).$$

Considering that T is a bijective map, if

$$\psi_t^M(x, y) = \psi_t^N(T(x), T(y))$$

for all $x, y \in M$ and all $t > 0$, then we have

$$h_t^M(x, y) = h_t^N(T(x), T(y)).$$

And by Proposition 2 in [Sun et al. 09], T is isometric.   □

This theorem implies that the isometric geometry can be completely determined by the Mexican hat wavelet. Hence, it is informative. It captures the geometry up to isometry.

Stableness   The construction of wavelets ψ_t from their Fourier transforms $\widehat{\psi}_t$ in spectral domain is stable. That is, perturbations of wavelets in frequency will not be amplified after the inverse Fourier transform.

Theorem 5.6 (Perturbation Theorem [Hammond et al. 11]) *For two functions in Fourier domain* $\widehat{\psi}_t(k)$, $\widehat{\psi}_t'(k)$, *if*

$$\sum_{k=0}^{\infty} \left(\widehat{\psi}_t(k) - \widehat{\psi}_t'(k) \right)^2 \leq \epsilon(t),$$

then for any $x \in M$,

$$\| \psi_t(x, y) - \psi_t'(x, y) \|^2 \leq \epsilon(t).$$

Proof: By the Parseval's Theorem 5.2, we have

$$
\begin{aligned}
\| \psi_t - \psi_t' \|^2 &= \langle (\psi_t - \psi_t'), (\psi_t - \psi_t') \rangle \\
&= \sum_{k=0}^{\infty} \left(\widehat{\psi}_t(k) - \widehat{\psi}_t'(k) \right)^2 \qquad (5.25) \\
&\leq \epsilon(t).
\end{aligned}
$$

□

This property means the computation of Mexican hat wavelet is stable under perturbations of its Fourier coefficients, and hence the Laplace-Beltrami eigenvalues.

5.5 Wavelet Transform

In this section, we analyze continuous wavelet transform (CWT) and discrete wavelet transform (DWT) of the Mexican hat wavelet from a Fourier perspective. Similar with the case in signal processing, we propose a method to compute convolutions by Fourier transform, which significantly improves the computational time of wavelet transforms, without reducing their accuracy.

5.5.1 Continuous Wavelet Transform

Theorem 5.7 (CWT of the Mexican hat wavelet) *For function $f(x) \in L^2(M)$, the CWT is given by*

$$\mathcal{W}_f(x,t) = \psi_t(x,y) * f(x), \tag{5.26}$$

and its Fourier transform is given by

$$\widehat{\mathcal{W}_f}(k,t) = \widehat{\psi}_t(k)\widehat{f}(k) \tag{5.27}$$

The inverse CWT is given by

$$f(x) = \mathcal{R}_f + \int_0^\infty \mathcal{W}_f(x,t)dt, \tag{5.28}$$

where

$$\mathcal{R}_f = \widehat{f}(0)\phi_0(x)$$

is a residual constant at the lowest frequency.

Proof: Eq. (5.26) is a direct consequence of the Convolution Theorem 5.1. For Eq. (5.28), we have

$$\begin{aligned}
\int_0^\infty \mathcal{W}_f(x,t)dt &= \int_0^\infty \sum_{k=0}^\infty \lambda_k e^{-\lambda_k t}\widehat{f}(k)\phi_k(x)dt \\
&= \sum_{k=1}^\infty \widehat{f}(k)\phi_k(x) \\
&= f(x) - \widehat{f}(0)\phi_0(x).
\end{aligned} \tag{5.29}$$

Hence, the statement is proved. □

In the CWT, x and t are localized in space and frequency, respectively, and $\mathcal{W}_f(x,t)$ is also called the wavelet coefficient of $f(x)$. It records detailed information of $f(x)$ at different scales. The convolution-based expression is analogous to the Fourier transform. Its time complexity is quadratic to the data size, since point-to-point function values are required. However, through its Fourier transform, we obtain a spectral expression of the CWT

$$\mathcal{W}_f(x,t) = \sum_{k=0}^{\infty} \lambda_k e^{-\lambda_k t} \widehat{f}(k)\phi_k(x), \tag{5.30}$$

which is only linear to the data size. This is consistent with the case in signal processing, that convolutions can be fast computed by the Fourier transform.

5.5.2 Discrete Wavelet Transform

Similarly, we can define the DWT, which will be useful in some applications. Assume we have a discrete sequence of samples $[t_0, t_1, \ldots, t_J]$. For the ease of formulation, we always let $t_0 = 0$, and $[t_1, t_2, \ldots, t_J]$ form a geometric sequence. We define the discrete Mexican hat wavelet as

$$\psi_{t_j}(x,y) = \frac{h_{t_{j-1}}(x,y) - h_{t_j}(x,y)}{t_j - t_{j-1}}, \tag{5.31}$$

and its Fourier transform is given by

$$\widehat{\psi}_{t_j}(x,y) = \frac{\widehat{h}_{t_{j-1}}(k) - \widehat{h}_{t_j}(k)}{t_j - t_{j-1}}, \tag{5.32}$$

for $j = 1, 2, \cdots, J$.

Theorem 5.8 (DWT of the Mexican hat wavelet) *For function $f(x) \in L^2(M)$, the DWT is given by*

$$\mathcal{W}_f(x,t_j) = \psi_{t_j}(x,y) * f(x), \tag{5.33}$$

and its Fourier transform is given by

$$\widehat{\mathcal{W}}_f(k,t_j) = \widehat{\psi}_{t_j}(k)\widehat{f}(k). \tag{5.34}$$

The inverse DWT is given by

$$f(x) = \mathcal{R}_f(x,t_J) + \sum_{j=1}^{J}(t_j - t_{j-1})\mathcal{W}_f(x,t_j), \tag{5.35}$$

where

$$\mathcal{R}_f(x,t_J) = h_{t_J}(x,y) * f(x)$$

is the residual function at the lowest-frequency sample t_J.

Proof: Eq. (5.33) is a direct consequence of the Convolution Theorem 5.1.
For Eq. (5.35), we have

$$\sum_{j=1}^{J}(t_j - t_{j-1})\mathcal{W}_f(x, t_j) = h_{t_0}(x, y) * f(x) - h_{t_J}(x, y) * f(x) \tag{5.36}$$
$$= f(x) - h_{t_J}(x, y) * f(x).$$

Hence, the statement is proved. □

We also obtain a spectral expression for the DWT:

$$\mathcal{W}_f(x, t_j) = \sum_{k=0}^{\infty} \frac{e^{-\lambda_k t_{j-1}} - e^{-\lambda_k t_j}}{t_j - t_{j-1}} \widehat{f}(k)\phi_k(x). \tag{5.37}$$

The aforementioned (continuous and discrete) inverse transforms imply
that we can, without loss of any information, recover a signal from its
wavelet coefficients and a residual function. Therefore, they can be used in
function approximation and analysis.

6

Anisotropic Wavelet

Isotropic diffusion refers to a diffusion process that transmits heat equally to all directions. It usually happens to a one-piece domain with uniform material and nature. A counterpart is anisotropic diffusion, which may behave differently at each direction. For example, heat diffusion from iron to wood is anisotropic at the joint. Anisotropic geometry diffusion and bilateral filter have been proposed to smooth bivariate data or general discretized surfaces. The anisotropic geometry diffusion [Desbrun et al. 00, Clarenz et al. 00, Hildebrandt and Polthier 04] is either complex or sensitive to boundary and scale changes. The bilateral filter [Jones et al. 03, Fleishman et al. 03, Su et al. 09] focuses on only local geometry information, which may also suffer from the problems of mesh quality and sampling. Moreover, they may result in degeneracy due to energy loss during the diffusion procedure.

These concepts can be easily applied to manifolds. Yet, an isotropic diffusion along a manifold can behave anisotropically in Euclidean directions. Most common diffusion processes are isotropic, for example the ones in previous chapters. This chapter introduces an anisotropic diffusion to get a glance of this field. The anisotropic diffusion is measured by a weighted heat kernel, with weights computed from normal-controlled coordinates. It derives anisotropic wavelets defined as differences of adjacent scales in diffusion space.

6.1 Normal-Controlled Coordinates

6.1.1 Definition

The normal-controlled coordinates (NCC) were introduced by [Wang et al. 11a], designed for anisotropic shape representation. Essentially, normal-controlled coordinates are variants of differential coordinates [Lipman et al. 04]. The NCC operator at point x is given by

$$\text{NCC}(x) = \frac{1}{\sqrt{A(x)}} \sum_{y \in N(x)} w_{xy}(x - y), \tag{6.1}$$

where $N(x)$ is a neighborhood of point x, $A(x)$ the vertex Voronoi area [Meyer et al. 02] of x, and w_{xy} is the NCC weight of points x and y. The inner product of NCC and the corresponding vertex normal is defined as normal signature (NS). It effectively represents local geometry information, such as feature size, convexity, and concavity.

The global geometry can then be decomposed into two sets of scalar data representing tangential and normal components, which collectively encode local parameterization (along local tangent plane) and local geometry information (perpendicular to local tangent plane), respectively. The local parameterization information is captured by the NCC operator coefficients w_{xy}, and the local geometry information is captured by the NS. Given any kind of evaluations on vertex normals, the local parameterization and geometry information are uniquely defined. Specifically in [Wang et al. 11a], they used the mean weighted by areas of adjacent triangles normals [Jin et al. 05]. Oftentimes, the local parameterization information can be considered to be unchanged, and hence only the operation of normal components is necessary. And the mesh geometry can be recovered linearly from coefficients of NCC operators and the updated NS with the aid of vertex normals.

For a point x, let $\mathbf{n}$ be the vertex normal, and P be its projection plane that can be any plane with the normal $\mathbf{n}$:

$$\mathbf{n}\cdot(x - y) = 0 \ , \text{ for } y \in P. \tag{6.2}$$

The rescaled normal weight $\widetilde{w}_{xy}$ in [Wang et al. 11a] is given by

$$\widetilde{w}_{xy} = \frac{\tan(\alpha_y/2) + \tan(\beta_y/2)}{\|x' - y'\|}, \tag{6.3}$$

where y' denotes the projection of y onto the plane P, and angles α_y and β_y are angles associated with y' in the projection plane (as shown in Fig. 6.1). Please note that the normal weight of y indirectly relates to y via projection. It is the mean value coordinate weight [Ju et al. 05] on the projection plane. The weight in Eq. (6.1) is given by normalization

$$\omega_{xy} = \frac{\widetilde{w}_{xy}}{\sum_{y \in N(x)} \widetilde{w}_{xy}} \tag{6.4}$$

6.1.2   Properties

In principle, normal-controlled coordinates are generalizations of differential coordinates. They share some common properties of conventional differential coordinates, such as easy implementation, efficiency, etc. Moreover, normal-controlled coordinates have their distinctive properties:

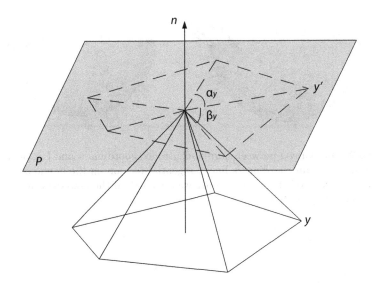

Figure 6.1. The construction of normal-controlled coordinates at a point.

- They are always parallel with the corresponding vertex normals with proof given below.

- They are scale-invariant. Assuming a mesh M is scaled to sM, the NCC are invariant under this scale:

$$\text{NCC}(sx) = \frac{1}{(s^2 A(x))^{1/2}} \sum_{y \in N(x)} w_{xy}(sx - sy) \qquad (6.5)$$
$$= \text{NCC}(x).$$

- They are well defined on boundary. For open meshes, the normals of boundary vertices can be well defined. Fig. 6.2 shows the Laplacian coordinates and NCC. The tangential tension appears in the Laplacian coordinates (a), while it completely disappears in NCC (b).

Proof (NCC Parallelism): Given any vertex x, its normal $\mathbf{n}$, its neighboring vertices $y \in N(x)$, and the normalized normal weights $\{w_{xy}\}$, we prove that $\text{NCC}(x)$ is parallel with the normal $\mathbf{n}$. That is,

$$\text{NCC}(x) = c \cdot \mathbf{n},$$

where c is a constant.

First, there exists a plane P^* with normal $\mathbf{n}$ passing through the point

$$\widetilde{x} = \sum_{y \in N(x)} w_{xy} y.$$

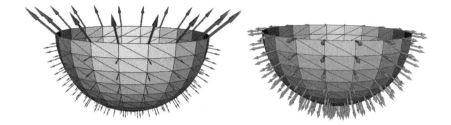

Figure 6.2. Comparison between normal-controlled coordinates and Laplacian co-ordinates. Laplacian coordinates have tangential tension on the boundary (left), while normal-controlled coordinates are well-defined on the boundary and insensitive to sampling (right).

Points $x^* \in P^*$ satisfy

$$P^* = x^* \cdot \mathbf{n} - \widetilde{x} \cdot \mathbf{n} = 0. \tag{6.6}$$

Second, let us project x and y onto the plane P^*, denoted as x^* and y^*, respectively. According to the reproduction property of mean value coordinates, we have

$$x^* = \sum_{y \in N(x)} w_{xy} y^* \tag{6.7}$$

$$y^* = y + c_y \cdot \mathbf{n}$$

where c_y is a constant. Obviously,

$$x - x^* = c_x \cdot \mathbf{n}, \tag{6.8}$$

where c_x is also a constant. Plug Eq. (6.7) into Eq. (6.8), we have

$$x - \sum_{y \in N(x)} w_{xy} y = c_x \cdot \mathbf{n} + \mathbf{n} \cdot \sum_{y \in N(x)} w_{xy} c_y. \tag{6.9}$$

It is easy to obtain the following equation from Eq. (6.6)

$$\sum_{y \in N(x)} w_{xy} c_y = 0.$$

Finally, we have

$$x - \sum_{y \in N(x)} w_{xy} y = c_x \cdot \mathbf{n}, \tag{6.10}$$

where the left-hand side of Eq. (6.10) is the NCC(x), and constant c_x is the magnitude of δ_i. Hence, the NCC are parallel with the corresponding vertex normals. $\square$

These properties are significant for mesh processing. They afford the stability, efficiency, and a wide range of applications of NCC.

6.2   Anisotropic Heat Kernel

It is well-known that the heat diffusion over manifold M is governed by the heat equation

$$\frac{\partial u(x,t)}{\partial t} = \Delta_M u(x,t), \tag{6.11}$$

where Δ_M is the Laplace-Beltrami operator. Solutions of this partial differential equation can be obtained by

$$u(x,t) = \int_M h_t(x,y)f(y)d\mu(y), \tag{6.12}$$

where $\mu(y)$ is the Riemannian volume of y, and $h_t(x,y)$ is the heat kernel between x and y at time t.

The commonly-used heat kernels are isotropic, which do not consider some geometry information. As a result, they lack control of heat diffusion and are sensitive to irregular sampling. A weighted linear finite element method (FEM) discretization of heat kernels was given in [Patanè and Falcidieno 10]. They discretized the Laplace-Beltrami operator in Eq. (6.11) by adding area weights. This discretization is stable to the sampling with the cost of losing symmetry. Despite the improvement, they still lack control of direction of heat diffusion.

At each vertex, heat diffuses through its connected paths (edges) as time goes by. Therefore, the edge weight plays an important role in thermal conductivity. A proper edge weight that can control the heat diffusion is preferred. Moreover, the area weights are used to compute eigen-system only to combat the problem of irregular sampling and symmetry of heat kernels. A weighted Laplace-Beltrami operator can be discretized by the edge-weighted Laplacian matrix [Patanè and Falcidieno 10, Reuter et al. 06]

$$L := B^{-1}L^e,$$

with

$$L^e(x,y) = \begin{cases} w(x,y) = e^{-\frac{\|NCC(x)-NCC(y)\|^2}{\sigma^2}}, & y \in N(x), \\ -\sum_{y \in N(x)} w(x,y), & x = y, \\ 0, & \text{otherwise,} \end{cases} \tag{6.13}$$

and

$$B(x,y) = \begin{cases} \frac{|t_r|+|t_s|}{12}, & y \in N(x), \\ \frac{\sum_{y \in N(x)} |t_y|}{6}, & x = y, \\ 0, & \text{otherwise,} \end{cases} \qquad (6.14)$$

where σ is a parameter proportional to the variance of NCC, t_r and t_s are the triangles that share the edge (x,y), and $|t_r|$ is the area of triangle t_r. Here, B is a mass matrix used to compensate the irregular sampling, and L^e is a weighted matrix that controls the tendency of heat diffusion (e.g., isotropic or anisotropic) using parameter σ.

The anisotropic eigen-system $\{\lambda_k^e, \phi_k^e\}$ of L is obtained from the generalized eigenvalue problem

$$L^e \phi_k^e = \lambda_k^e B \phi_k^e. \qquad (6.15)$$

Using the eigenfunctions, edge-weighted heat kernels can be analytically written as

$$h_t^e(x,y) = \sum_{k=0}^{\infty} e^{-\lambda_k^e t} \phi_k^e(x) \phi_k^e(y). \qquad (6.16)$$

The anisotropic heat kernels are determined by the discretized edge-weighted Laplace-Beltrami operator, which incorporates more geometry. One can control the heat diffusion easily by adjusting the edge weight $w(x,y)$. Intuitively, the heat diffuses faster along the prominent parts, but rather slow when cutting across them, such as sharp edges. Fig. 6.3 (a) and (b) show the difference between commonly-used heat kernels and the edge-weighted heat kernels. It clearly illustrates that edge-weighted heat kernels are more aware of local geometry. Moreover, edge-weighted heat kernels are robust with respect to noise and perturbation, as shown in Fig. 6.3 (c).

The discretization of the heat kernel introduced above can be easily refined to be scale invariant using the method in [Patanè and Falcidieno 10], if scale changes are encountered. Specifically, assume manifold M is scaled to sM. The matrix L^e is unchanged, the mass matrix B becomes $s^2 B$, and the eigen-system turns into $\{\lambda_k^e/s^2, \phi_k^e/s\}$. The edge-weighted heat kernels can be modified to

$$h_t^{e'}(x,y) = \beta \sum_{k=0}^{\infty} e^{-\lambda_k' t} \phi_k^e(x) \phi_k^e(y), \qquad (6.17)$$

with $\beta = \frac{B(x,x)+B(y,y)}{2}$, and $\lambda_k' = \frac{\lambda_k^e}{\lambda_1^e}$, which is scale invariant.

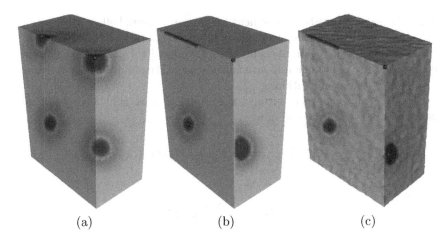

(a)   (b)   (c)

Figure 6.3. Comparison between the commonly-used heat kernels (a), edge-weighted heat kernels without noise (b) and with noise (c). The heat kernels of four different vertices on a cuboid are shown, respectively.

6.3 Anisotropic Diffusion

Given an initial heat field $f_0(x)$ defined by, for example the normal signature of M, we can diffuse it isotropically or anisotropically using edge-weighted heat kernels. We exploit heat kernel convolution in a local region to ensure the heat conservation and accelerate the computation.

6.3.1 One-Step Diffusion

The multi-scale property of heat kernels implies that for small value of time t, heat kernels can be well approximated by the ones within a small geodesic neighborhood of vertex x [Sun et al. 09, Ovsjanikov et al. 10]. An explicit relationship between time and the size of diffusion region was given in [Grigor'yan 99].

The heat kernel $h_t(x, \cdot)$ can be viewed as the transition density function of the Brownian motion, and is determined by the geodesic ball

$$B(x, O(\sqrt{c_n t \log t})).$$

Mémoli further justified the interpretation of time as a geometric scale from the view point of homogenization of partial differential equations [Mémoli 09]. Therefore, it is justifiable and acceptable to define a heat region around x as

$$\Omega_t^x = \{y \mid h_t^e(x, y) > \epsilon(h)\}, \tag{6.18}$$

where t_h is some threshold. The heat region can be found efficiently by taking advantage of the nearest neighbor method [Hastie and Tibshirani 96]. Moreover, the heat region of each vertex for a fixed time t only needs to be computed once. The region is invariant during diffusion using our iterative local heat kernel convolution.

Considering a meshed surface, the updated heat field after one-time convolution in the heat region is

$$f(x,t) = \frac{1}{C^*(x)} \sum_{y \in \Omega_t^x} h_t^e(x,y) f_0(y), \qquad (6.19)$$

where

$$C^*(x) = \sum_{y \in \Omega_t^x} h_t^e(x,y)$$

is for normalization, and $f_0(y)$ is an initial field. This can be written in matrix form

$$F_1 = H_t F_0, \qquad (6.20)$$

where

$$F_1 = [f(x_1,t), \ldots, f(x_n,t)]^T$$

$$F_0 = [f_0(x_1), \ldots, f_0(x_n)]^T,$$

and H_t is a sparse diffusion matrix with elements

$$H_t(x,y) = \begin{cases} \frac{1}{C^*(x)} h_t^e(x,y), & \text{if } y \in \Omega_t^x \\ 0, & \text{otherwise.} \end{cases} \qquad (6.21)$$

6.3.2 Semigroup Transmission

According to the semigroup identity of heat kernels, the heat diffusion can be obtained by using the local heat kernel convolution iteratively. By fixing a time t for one-step diffusion in local heat region, the heat field after j steps is

$$F_j = H_t F_{j-1} = \cdots H_t^{j-1} F_1 = H_t^j F_0. \qquad (6.22)$$

When $j = 0$, the heat field is the initial field. As k increases, the heat diffuses to a larger region, and more global behavior results from the convolution. The diffusion turns into matrix-vector multiplication. Since H_t is a sparse matrix, the computation is very fast.

There are two important parameters σ and j in our approach. The parameter σ in Eq. (6.13) controls the tendency of heat diffusion in some sense. When the data has noise, σ should be larger, and vice versa. On the other hand, if the preservation of sharp features is important, σ should be small so that neighboring NCC deviating far from the current one makes very small contributions to itself. The parameter k in Eq. (6.22) controls the diffusion level. In a nutshell, the bigger the j is, the larger region the heat will diffuse to. The other parameters, such as time t and the number of eigenvalues m, are robust in our approach. To make the time t meaningful and stable for different surfaces, we rescale the surface such that the total area is equal to the number of vertices. Thus, $t = 1$ results in an average influence region of roughly 1-ring size [Vaxman et al. 10], i.e., one step in random walk.

6.4 Anisotropic Wavelet

Similar to the diffusion wavelets (Chapter 4) and the Mexican hat wavelet (Chapter 5), we can formulate anisotropic wavelet, which can also be used for space-frequency processing.

6.4.1 Continuous Anisotropic Wavelet

Considering the edge weighted heat kernel $h_t^e(x, y)$, one can define the anisotropic Mexican-hat-type wavelet as

$$\psi_t^e(x, y) = -\frac{\partial h_t^e(x, y)}{\partial t} = \sum_{k=0}^{\infty} \lambda_k^e e^{-\lambda_k^e t} \phi_k^e(x) \phi^e(y), \qquad (6.23)$$

with an analytical expression in Fourier domain

$$\widehat{\psi_t^e} = \lambda_k^e e^{-\lambda_k^e t}. \qquad (6.24)$$

It is similar with the regular Mexican hat wavelet $\psi_t(x, y)$, and they share many properties.

Analogously, one can define continuous wavelet transform (CWT) of the anisotropic Mexican hat wavelet. For a given function $f(x) \in L^2(M)$, its continuous wavelet transform is defined as the inner product of wavelet $\psi_t^e(x, y)$ and function $f(x)$, given by

$$\mathcal{W}_f^e(x, t) = \langle \psi_t^e(x, y), f(y) \rangle. \qquad (6.25)$$

As we know, in Fourier domain, this is simple a dot product

$$\widehat{\mathcal{W}_f^e}(k, t) = \widehat{\psi_t^e}(k) \widehat{f}(k). \qquad (6.26)$$

Consequently, the inverse transform is given by

$$f(x) = \mathcal{R}_f^e + \int_0^{\infty} \mathcal{W}_f^e(x, t) dt, \qquad (6.27)$$

where
$$\mathcal{R}_f^e = \widehat{f}(0)\phi_0^e(x)$$
is a residual constant at the lowest frequency.

The anisotropic wavelet transform has exactly the same format with the regular one in Chapter 5. Therefore, it can directly fit in all algorithms and applications of the regular Mexican hat wavelet.

6.4.2   Discrete Anisotropic Wavelet

The design of discrete anisotropic wavelet is based on the semigroup diffusion built by the one-time step diffusion matrix H_t. For a linear sequence $\{jt\}$ with
$$j = 0, 1, \ldots, J,$$
the discrete anisotropic wavelet matrix Ψ_j is defined as

$$\Psi_j = \frac{H_t^{j-1} - H_t^j}{t}, \qquad (6.28)$$

where t is a fixed small time for one-step local diffusion. When $j = 0$, the diffusion matrix becomes the identity matrix

$$H_t^0 = I. \qquad (6.29)$$

Considering diffusion is conducted slowly along t, one can also take a geometry sequence $\{2^j t\}$, like the diffusion wavelets (Chapter 4) and the discrete Mexican hat wavelet (Chapter 5). The sequence of diffusion matrix is now $\{H_t^{2^j}\}$. Accordingly, the discrete anisotropic wavelet matrix Ψ_j is then given by

$$\Psi_j = \frac{H_t^{2^{j-1}} - H_t^{2^j}}{2^{j-1} t}. \qquad (6.30)$$

Since in the diffusion matrix H_t, edge weighted heat kernels are normalized, the discrete anisotropic wavelet satisfies the zero-mean property and the admissibility property.

Analogously, one can define discrete wavelet transform (DWT) of the discrete anisotropic wavelet. For a given function $f(x) \in L^2(M)$, its discrete wavelet transform is defined as the product of wavelet matrix Ψ_j and function vector F, given by

$$\mathcal{W}_F(j) = \Psi_j F. \qquad (6.31)$$

This is actually the difference between two adjacent scales.

For the linear sequence $\{jt\}$, the inverse transform is given by

$$f(x) = H_t^J + t \sum_{j=1}^{J} \mathcal{W}_F(j). \qquad (6.32)$$

And for the geometric sequence $2^j t$, the inverse transform is given by

$$f(x) = H_t^{2^J} + t \sum_{j=1}^{J} 2^{j-1} \mathcal{W}_F(j). \tag{6.33}$$

Readers can prove the inverse transform by simply expanding the sum and replacing wavelet transform terms by their definitions.

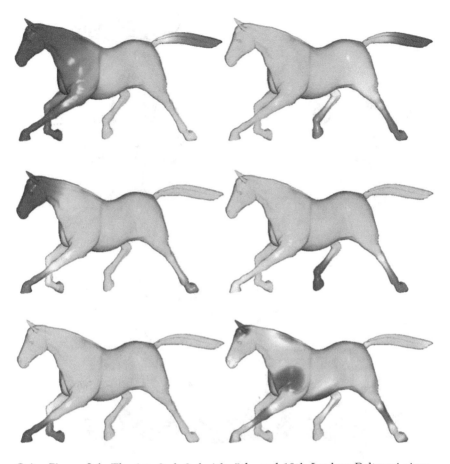

Color Figure 3.1. The 1st, 2nd, 3rd, 4th, 5th, and 10th Laplace-Beltrami eigenfunctions of a closed shape, rendered by coded color.

$\varphi_{4,x}$ $\quad\quad\quad$ $\psi_{4,x}$

Max

0

Min

Color Figure 4.2. Rendering a scaling function and a wavelet of the admissible diffusion wavelets on a meshed surface by color coding.

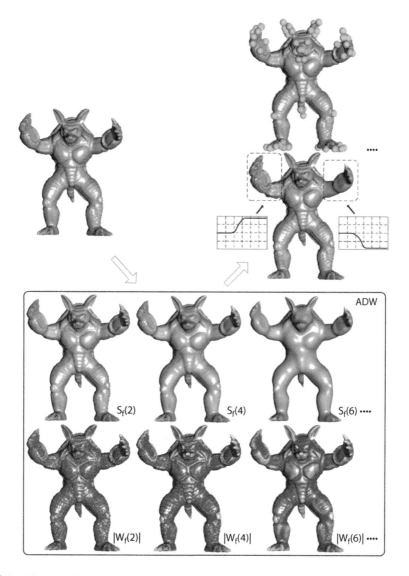

Color Figure 4.6. The application framework of the admissible diffusion wavelets.

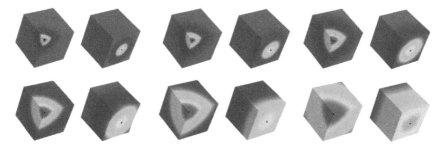

Color Figure 5.2. Mexican hat wavelets on a box at $t = 10, 20, 40, 80, 160, 320$. The centers of functions are marked by black balls. The Mexican hat wavelet behaves differently at a corner point and a planar point.

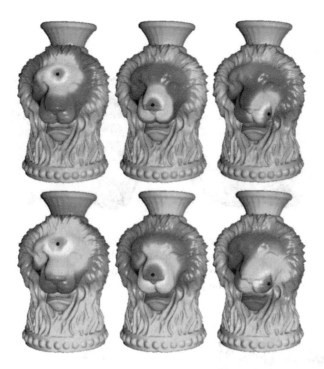

Color Figure 5.3. Some Mexican hat wavelets on a 2D manifold visualized by color plots, at $t = 50$ (top) and $t = 100$ (bottom). Wavelets are attenuated and oscillating on the manifold.

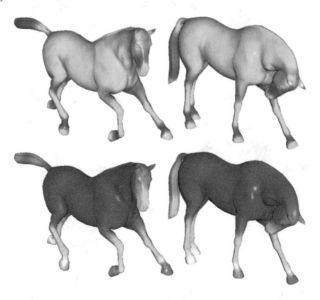

Color Figure 5.4. Visualizing $\psi_t(x, x)$ at $t = 80$ (Top) and $t = 640$ (Bottom) on deformable shapes. It is relevant to shape geometry, and invariant to isometric deformation.

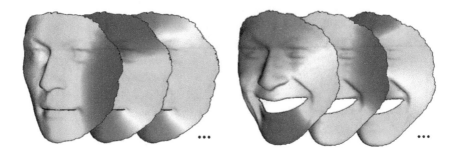

Color Figure 9.1. Some Laplace-Beltrami eigenfunctions of two partial shapes undergoing deformation and boundary changes. The two shapes share most common part of the object. However, the orders of their eigenfunctions vary due to deformation and boundary changes.

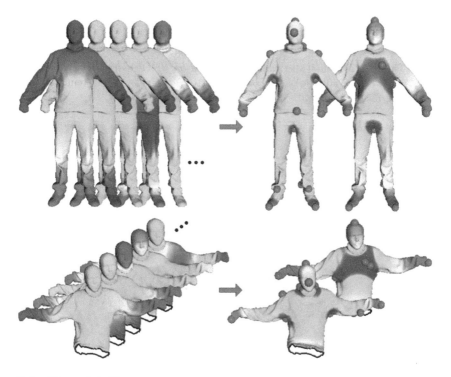

Color Figure 9.2. Shape representation methods. Two shapes that are partially overlapped with deformation have different Laplace-Beltrami eigenfunctions (left) but similar heat kernel signatures (right).

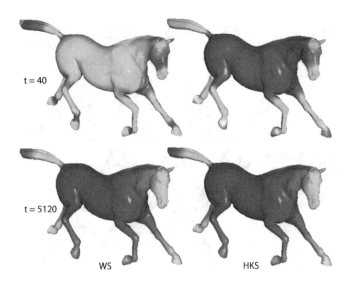

Color Figure 9.6. Comparison of wavelet signature (WS) and heat kernel signature (HKS) as shape representation. While both of the two are similar at large time, wavelet signature is more distinguishable at small time.

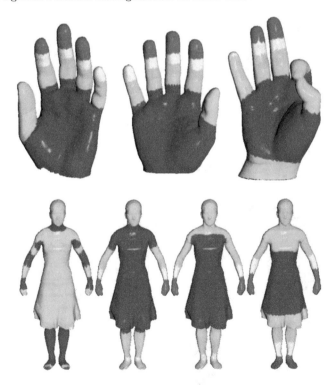

Color Figure 9.7. Segmentation by clustering wavelet signatures. Top: results of hand models with different topologies: open surface, closed surface, and genus-1 surface. Bottom: multiscale segmentation at different times.

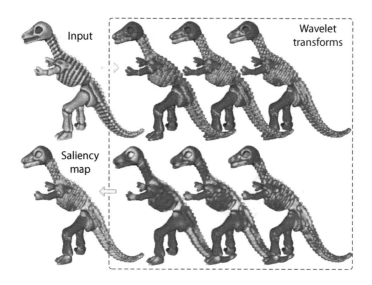

Color Figure 11.1. A saliency map is the nonlinear sum of its wavelet coefficients.

Color Figure 11.4. Saliency visualization and feature detection on point clouds.

Color Figure 11.7. Mexican hat wavelet transform of vertex coordinates at different scales.

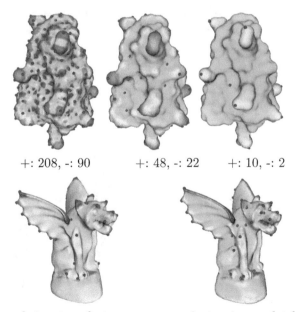

+: 208, -: 90   +: 48, -: 22   +: 10, -: 2

heat-kernel-signature features   anisotropic-wavelet features

Color Figure 11.9. Top: Anisotropic features detected at different scales on the Blob. The convex (+) and concave (-) features are highlighted in blue and red balls, respectively. Bottom: Comparison of heat-kernel-signature features (left) and anisotropic-wavelet features (right) detected on the Gargoyle.

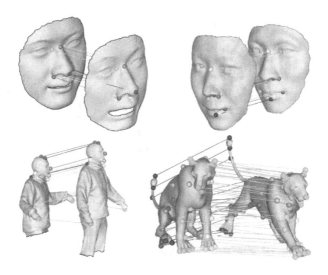

Color Figure 12.2. Spectral matching results from [Hou and Qin 12b].

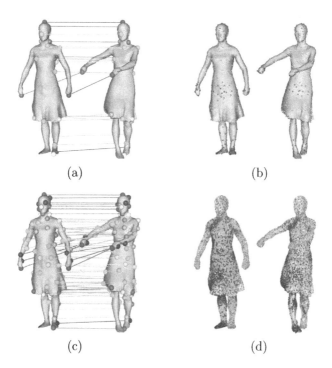

(a)

(b)

(c)

(d)

Color Figure 12.10. Major steps in the hierarchical registration algorithm. The blue shape is the source and the red one is the target. We use a three level hierarchy in this example. (a) Initial feature correspondences; (b) Coarse registration result (Third level); (c) Expanded feature correspondences (Third level); (d) Final registration result.

7

Wavelet Generation

The previous three chapters demonstrate different designs of wavelets on 3D shapes, from which we can conclude schemes to generate new wavelets. Two key techniques are heat diffusion and spectral theory. This chapter presents wavelet generation in two directions: to volumetric data and to general manifold wavelets.

7.1 Volume Wavelets

Volume is an important data format frequently used in medical imaging. In general, it is a 3D Euclidean space where classical wavelets can apply. However, certain difficulties prevail in this direction. The challenges are prompted by the fact that volumetric images typically have much more complex inner structure, and are frequently accompanied by non-rigid deformation with higher degrees of freedom (DOFs) in orientation. The deformable nature of volumetric images and their spatial flexibility have prevented the widespread penetration of 3D wavelets in volumetric medical image processing and analysis. Directly applying 3D wavelets does not respect different materials in the volume. It is very sensitive to noise, with unavoidable ambiguities for the ones with low contrast or being poorly localized nearby an edge.

In the work of [Li et al. 12], a material-aware construction was proposed, which can compute anisotropic wavelets on volumetric images.

7.1.1 Differential Geometry of Volumetric Images

The rich differential geometry theory offers an elegant method by treating an image as a differentiable surface [Zhang et al. 10]. A volumetric image is formulated as:

$$I(p) : \Omega \to \mathbb{R}^n \tag{7.1}$$

where $\Omega \in \mathbb{R}^3$ is the image domain, n is the number of image channels, and p represents a voxel. Thus, $I(p_1) - I(p_2)$ represents the intensity/color difference between p_1 and p_2.

The structure tensor (i.e., the second-moment matrix) is a matrix derived from function gradient. It locally characterizes the predominant directions of the gradient, and how those directions are related to each other. As the simplest tensor, structure tensor is derived from first-order differential analysis, which can locally characterize the predominant directions of material changes and how those directions are related to each other. The structure tensor, as a measurement for edges and their orientation, has been introduced to nicely inherit this feature through image derivatives. The classical structure tensor [Bigun et al. 91] was constructed by way of the tensor product of the smoothed image gradient, which has been widely accepted in texture analysis applications [Wang and Vemuri 04]. Malcolm [Malcolm et al. 07] generalized the tensor method to segment images by taking the Riemannian geometry of the tensor space into account. However, compared with Gabor filters, the structure tensor used in [Malcolm et al. 07] only reflects the orientation information at a single scale and fails to discriminate textures which are varying across different scales. Recently, the multiscale structure tensor proposed in [Scheunders 02, Han et al. 09] has demonstrated successful applications in image fusion and enhancement, and noise filtering.

Let $g(p) = (I_x(p), I_y(p), I_z(p))$ be the normalized gradient of voxel p, where subscripts x, y, z are the three dimensions in $\mathbb{R}^3$. One can define a local structure tensor of voxel p as:

$$T_S(p) = \begin{bmatrix} (I_x(p))^2 & I_x(p)I_y(p) & I_x(p)I_z(p) \\ I_y(p)I_x(p) & (I_y(p))^2 & I_y(p)I_z(p) \\ I_z(p)I_x(p) & I_z(p)I_y(p) & (I_z(p))^2 \end{bmatrix}. \tag{7.2}$$

For a normal vector $\mathbf{n}$, $\mathbf{n}^T T_S(p)\mathbf{n}$ measures the rate of change of the image along $\mathbf{n}$ and is referred to as the squared local contrast. It reaches its maximum when $\mathbf{n}$ is in the direction of the eigenvector corresponding to the largest eigenvalue of $T_S(p)$, while this eigenvector matches with the gradient direction g. The smoothed version of $T_S(p)$ allows the integration of the first-order information from the neighborhood and gives rise to a more stable numerical derivation:

$$T_S(p)_\sigma = G_\sigma * T_S(p), \tag{7.3}$$

where G_σ is the Gaussian kernel with standard deviation σ.

The structure tensor $T_S(p)$ or $T_S(p)_\sigma$ is a positive semi-definite tensor of the second-order; it can be diagonalized by eigenvalues

$$\lambda_1 > \lambda_2 > \lambda_3 \geq 0$$

and reformulated by a spectral representation:

$$T_S(p) = \lambda_1 \mathbf{e}_1 \mathbf{e}_1^T + \lambda_2 \mathbf{e}_2 \mathbf{e}_2^T + \lambda_3 \mathbf{e}_3 \mathbf{e}_3^T, \tag{7.4}$$

where $\mathbf{e}_k$ is the corresponding eigenvector of λ_k.

The Hessian matrix H of volumetric image describes the local second-order structure around each voxel and intuitively depicts how the normal (i.e., gradient) of an iso-surface changes. This symmetric matrix comprises second-order partial derivatives:

$$H(p) = \begin{bmatrix} I_{xx}(p) & I_{xy}(p) & I_{xz}(p) \\ I_{yx}(p) & I_{yy}(p) & I_{yz}(p) \\ I_{zx}(p) & I_{zy}(p) & I_{zz}(p) \end{bmatrix}. \tag{7.5}$$

The Hessian matrix H also has real-valued eigenvalues and corresponding eigenvectors

$$\lambda_1 > \lambda_2 > \lambda_3 \geq 0.$$

The eigenvectors point in the directions of the principal curvatures and the eigenvalues correspond to the curvature along those directions. Intuitively speaking, the directions corresponding to maximal eigenvalue of H should represent the most direct change from one material to adjacent neighboring material, while the directions corresponding to minimal eigenvalue show the material interface and how such material flows along the interface.

7.1.2 Diffusion Tensor Space

As already demonstrated in previous work, the proper definition of tensor space over a scalar image will be a key to local geometry structure analysis and subsequent image processing. Considering first-order derivatives cannot fully grasp the local geometric differential property, [Li et al. 12] employed Hessian eigen-system to define the local diffusion tensor, which facilitates the description of the second-order structure around each voxel and intuitively depicts how the normal of a surface changes.

Since directly imitating the definition of structure tensor will result in rapid diffusion when cutting across sharp material boundaries and slow diffusion when traveling along them, they reformulate an anisotropic diffusion tensor by a spectral representation as:

$$T_D(p) = \widetilde{\lambda}_1 e_1 e_1^T + \widetilde{\lambda}_2 e_2 e_2^T + \widetilde{\lambda}_3 e_3 e_3^T, \tag{7.6}$$

where

$$\widetilde{\lambda}_k = e^{-\frac{\lambda_k}{\sigma_d}}, \, k = 1, 2, 3, \tag{7.7}$$

with a diffusion parameter σ_d that controls the diffusion velocities. Here λ_k and e_k are the corresponding eigenvalue and eigenvector of the Hessian matrix.

Intuitively speaking, this constructs an ellipsoid at each voxel that encodes the direction and velocity of diffusion [Wang et al. 11b]. According to the theory of Rayleigh quotient [Horn and Johnson 85], the diffusion velocity from voxel p along e can be viewed as the length of the vector projection onto the ellipsoid, which is expressed as

$$vel(p, e) = \frac{e^T T_D(p) e}{e^T e}. \tag{7.8}$$

For a voxel inside a blob, all of its diffusion directions are principal diffusion directions, since all the diffusion velocities are equal. For a voxel on a boundary surface, all the directions aligning with its tangent plane are principal diffusion directions. For a voxel on a sharp edge, the direction along the edge is principal diffusion direction. For an isolated noise voxel, it will have no principal diffusion directions, as the velocities along all the directions are extremely small.

7.1.3   Anisotropic Volume Diffusion

Apart from the discontinuity due to surfaces and boundaries in volumetric images, they still have various directions due to their intrinsic material structure. The visual perception research has indicated that the cells having directional selectivity are found in the retinas and visual cortices of the entire major vertebrate classes; thus naively using the anisotropic kernel will naturally give rise to directional information loss among nearby regions without having evident clues on material structure. In order to respect the direction information embedded in the local structure during feature extraction, [Li et al. 12] shows a construction of anisotropic diffusion on volumetric images, which is derived from the diffusion tensor and bilateral filter.

The anisotropic diffusion considers both the material continuity and the photometric similarity. It prefers nearby values to distant values in both spatial and material domain (diffusion tensor space), which can relax the underlying assumption that images always vary slowly in the vicinity. Given two neighboring voxels located at p and q, a distance in diffusion tensor space is given by [Li et al. 12]

$$d_{T_D}(p, q) = e^{-(p-q)^T (w_{pq}(T_D(p) + T_D(q))^{-1}(p-q)} \tag{7.9}$$

where w_{pq} is introduced to amend the gradient, which changes in response to the intensity change of voxels. In fact, $T_D(p) + T_D(q)$ describes the diffusivity and controls the diffusion directions and velocities, and w_{pq} respects the intensity variance between neighboring voxels. The anisotropic diffusion is a convolution of a bilateral kernel and a volumetric image, given by

$$I(p)_{\sigma_s, \sigma_k} = \frac{1}{W_p} \sum_{q \in N(p)} G_{\sigma_s}(p - q) G_{\sigma_k}(d_{T_D}(p, q)) I(q). \tag{7.10}$$

where W_p is a normalization factor, $G_\sigma(p)$ is a spatial Gaussian kernel, and σ_k is a control parameter being set to the inverse of the maximal eigenvalues of diffusion matrices $T_D(p)$ and $T_D(q)$.

7.1.4 Anisotropic Volume Wavelet

With the built-in capability to faithfully respect material structure, one can use the anisotropic diffusion to compute wavelet transforms of a volumetric image. It actually decomposes a volumetric image into an approximation sub-band and a detail sub-band. However, only one-level decomposition is not enough to extract the feature information since images may be noisy and objects inherently comprise different details changing as a function of the observation scale. A mutiscale diffusion is introduced in [Li et al. 12], given by

$$I^{j+1}(p, \sigma_s) = \frac{1}{W_p} \sum_{q \in N(p)} G_{\sigma_s}(p - q)) G_{\sigma_k}(d_{T_D}^m(p, q)) I^j(q), \qquad (7.11)$$

where j represents the j-th level of the decomposition and W_p has the same meaning as Eq. (7.10). In the interest of time performance, distance $d_{T_D}^j(p, q)$ is updated only in response to the change of w_{pq} while preserving the $T_D(p) + T_D(q)$ according to Eq. (7.9).

In implementation, it is iterated over the approximate sub-bands and only the one-ring neighbors of each voxel are considered in each iteration. The Gaussian function embedded in the anisotropic diffusion is a unique distribution optimizing localization in both the spatial and frequency domains. It indicates that the diffusion can be applied within local support with minimum aliasing errors.

After $j + 1$ iterations, the approximate sub-band corresponding to a certain scale can be obtained. Given an initial parameter σ, the j-th detail sub-band, i.e. the wavelet transform, is given by the difference between the neighboring approximate sub-band as

$$\mathcal{W}_I(p, j) = I^{j+1}(p, \sigma) - I^j(p, \sigma). \qquad (7.12)$$

Fig.7.1 shows that a volumetric image is decomposed to an approximate sub-band and some detail sub-bands by the anisotropic volume wavelet.

The above construction is an extension of the anisotropic wavelet in Chapter 6. The scheme is to first build a scale space of an input function by some diffusion kernel, and then deduce wavelet transform as the difference of two adjacent scales. It is in spirit a wavelet because it captures details at different scales by the sub-bands. Like the admissible diffusion wavelets in Chapter 4 and the anisotropic wavelet in Chapter 6, the scale corresponds to the frequency. Small scales correspond to high frequencies while large scales

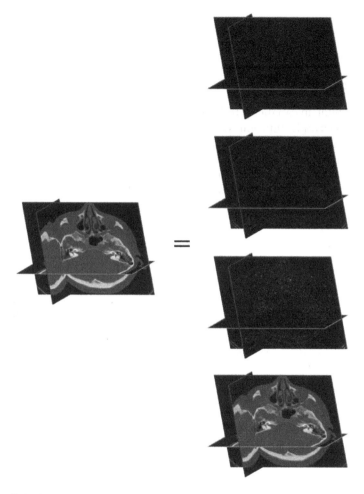

Figure 7.1. A volumetric image is decomposed to an approximate sub-band and some detailed sub-bands by the anisotropic volume wavelet.

correspond to low frequencies. Hence, the wavelet also has localization in both space and frequency.

7.2   Manifold Wavelet Generalization

Inspired by the spectral graph wavelets, one can define general manifold wavelets by some kernels in manifold Fourier domain. As shown in Chapter 5, functions without closed-form expressions on manifolds may have analytical expressions in Fourier domain. This greatly facilitates the wavelet design on manifolds.

7.2.1   Generalization Scheme

In [Antoine et al. 10], a similar design for wavelet generation on graphs was proposed. For the ease of understanding, we will use the same notations. This wavelet design is also based on the graph Laplacian Δ_G and its eigen-system $\{\lambda_k, \phi_k\}$, which can be used to define a class of operator-valued functions via their spectral decomposition. This is called the Borel functional calculus [Antoine et al. 10]. For a wavelet kernel $g \in L^2$, the operator-valued function T_g is defined by

$$T_g f = \sum_{k=0}^{N-1} g(\lambda_k) \widehat{f}_k \phi_k, \qquad (7.13)$$

where

$$\widehat{f}(k) = \langle \phi_k(x), f(x) \rangle$$

denotes the graph Fourier coefficient of f. In a broader sense, we call $g(\lambda_k)$ the Fourier Transform of the operator T_g, and $g(\lambda_k)\widehat{f}(k)$ the Fourier coefficient of the function $T_g f$.

It is natural to dilate the operator-valued function $T_g(\Delta_G)$ by scaling its Fourier coefficients with t:

$$(T_g(t\Delta_G)f)(x) = \sum_{k=0}^{N-1} g(t\lambda_k)\widehat{f}(k)\phi_k(x). \qquad (7.14)$$

Similar ideas have been used in the needlets [Narcowich et al. 96] and spherical wavelets [Holschneider 96, Freeden and Windheuser 97]. The graph wavelets indexed by scale and location are sampled from the wavelet operator by impulse function,

$$\psi_t(x, y) := T_g(t\Delta_G(x, y))\delta(x, y). \qquad (7.15)$$

Antoine et al. defined the graph wavelet transform of a function f as the inner product [Antoine et al. 10]

$$\mathcal{W}_f(x, t) = \langle \psi_t(x, y), f(y) \rangle. \qquad (7.16)$$

And the inverse transform is given by

$$f(x) = c_\psi^{-1} \sum_i \int \frac{dt}{t^\beta} \mathcal{W}_f(x, t) \psi_t(x, y), \qquad (7.17)$$

where β is an arbitrary weight.

This approach can be easily lifted to manifolds for generating continuous wavelets with generator g. The lifting is based on the Fourier transform on manifold harmonics. Wavelets are generated as

$$\psi_t(x, y) = \sum_{k=0}^{\infty} g(t\lambda_k)\phi_k(x)\phi_k(y), \qquad (7.18)$$

where $\{\lambda_k, \phi_k(x)\}_{k=0}^{\infty}$ compose eigen-system of the Laplace-Beltrami operator. Please note that the number of eigenvalues goes to infinity, because it is on differentiable manifolds. It is easy to see that the generator $g(t\lambda_k)$ is the Fourier coefficient of the wavelet $\psi_t(x, y)$. For a square integrable function $f(x) \in L^2(M)$, its transform of generated wavelets is given by

$$\mathcal{W}_f(x, t) = \langle \psi_t(x, y), f(y) \rangle = \sum_{k=0}^{\infty} g(t\lambda_k)\widehat{f}(k)\phi_k(x), \qquad (7.19)$$

where $\widehat{f}(k)$ is now the Fourier coefficient of $f(x)$ by manifold harmonics. This is the inverse Fourier transform of coefficients $g(t\lambda_k)\widehat{f}(k)$. The generated wavelets are defined by the wavelet operator, with similar numerical expansion using Laplace-Beltrami eigenfunctions.

In a recent work [Kim et al. 13], this scheme is applied to 3D mesh segmentation and alignment. It considers the self-effect of the wavelet kernel localized on itself $\psi_t(x, x)$, and defines a wavelet kernel descriptor with normalization at each frequency t,

$$\text{WKD}(x, t) = \frac{\psi_t(x, x) - \min_y \psi_t(y, y)}{\max_y \psi_t(y, y) - \min_y \psi_t(y, y)}. \qquad (7.20)$$

7.2.2   Wavelet Prototypes

In principle, defining wavelets on manifolds is extremely challenging. However, this new approach affords a systematic way to define general wavelets in spectral domain. It is then possible to define functions in Fourier domain, and transform them back to manifolds. We shall discuss some meaningful rules to design the wavelet generator. The generator g is the Fourier transform of wavelet $\psi_t(x, y)$. Let us use $\widehat{\psi}_t(k)$ to represent the Fourier transform of wavelet $\psi_t(x, y)$.

First, the wavelet should have vanishing zero-frequency to satisfy the admissibility condition. It means

$$\widehat{\psi}(0) = 0,$$

and the wavelet has zero-mean, which implies

$$\int_M \psi_t(x, y) d\mu(y) = 0.$$

This ensures the non-trivial wavelet to oscillate on the manifold. It can be achieved by forcing non-differentiable functions to be zero at $k = 0$, or a power function of λ_k with non-zero power, since $\lambda_k = 0$ for the Neumann Laplace-Beltrami operator.

Second, the wavelet should be dilated by t. This means the wavelet is multiscale with different scales in space. In diffusion wavelets, the dilation is achieved by dyadic powers of a diffusion operator. In the generated wavelets, we can use t as a scale applying to λ_k for dilation, resulting in $\lambda_k t$.

Third, the wavelet should be band-pass, so that it can capture information from the signal at different bands. This can be solved by an exponential function of $\lambda_k t$, which decreases rapidly in frequency domain.

Based on these rules, we propose some prototypes of newly generated wavelets formulated using their Fourier transforms. These formulations are derived similar to the ones in Euclidean metric, by means of transferring functions in spectral domain. Interested readers are encouraged to explore them with more details by themselves.

- Mexican hat wavelet:

$$\widehat{\psi}_t(k) = \lambda_k e^{-\lambda_k t}. \tag{7.21}$$

It has been studied in Chapter 5.

- Morlet-type wavelet:

$$\widehat{\psi}_t(k) = e^{-|\lambda_k - \lambda_K|t}, \tag{7.22}$$

where λ_K is a large frequency, and

$$\widehat{\psi}(0, t) = e^{-\lambda_K t} \to 0.$$

It is a Gaussian being translated to λ_K in the frequency domain.

- Gauss-type wavelet:

$$\widehat{\psi}_t(k) = \sqrt{\lambda_k t}\, e^{-\lambda_k t}. \tag{7.23}$$

Consider $\sqrt{\lambda_k t}$ corresponds to the frequency in Euclidean domain; it is similar to the Gaussian wavelet.

- Marr-type wavelet:
$$\widehat{\psi_t}(k) = \lambda_k t e^{-\lambda_k t}. \tag{7.24}$$

- High-order wavelet:

$$\widehat{\psi_t}(k) = (\lambda_k t)^p e^{-\lambda_k t}, \tag{7.25}$$

with constant $p > 1$. It is a generalization of the Marr-type wavelet.

- Shannon-type wavelet:

$$\widehat{\psi_t}(k) = e^{-\lambda_k t} \chi(\sqrt{\lambda_k t}), \tag{7.26}$$

with

$$\chi(\omega) = \begin{cases} 1, & \pi < \omega < 2\pi \\ 0, & \text{otherwise} \end{cases}.$$

- Littlewood-Paley-type wavelet:

$$\widehat{\psi_t}(k) = (2\pi)^{-1/2} \chi(\sqrt{\lambda_k t}). \tag{7.27}$$

The prototypes wavelets are icing on the cake for designing manifold wavelets in Fourier domain. Investigating these generated wavelets is an effective practice to study wavelet design on manifolds. Following this scheme, many more wavelets can be designed and serve for different purposes.

Besides, it may also lead to new research directions in computer graphics and other related areas. The spectral graph wavelets are fresh results in mathematics. Valuable topics are waiting to be explored for followers in computer science areas.

Part II

Applications

8

Implementation

This chapter reveals implementation details of fundamental operations for computing diffusion driven wavelets on 3D shapes, including discrete Laplace-Beltrami operator, generalized eigenvalue problem, and matrix power, through which readers should be able to implement most diffusion-related techniques on 3D shapes.

Among current research results, three schemes were used for implementing diffusion on 3D shapes: expansion on Laplace-Beltrami eigenfunctions [Sun et al. 09, Hou and Qin 12a], matrix power of a local operator [Hou and Qin 10, Hou and Qin 13], and a compound of the first two [Vaxman et al. 10, Wang et al. 11a]. All of the schemes are utilized in this book. Typically, the first scheme requires computing discrete Laplace-Beltrami operator on a 3D shape, and solving a generalized eigenvalue problem. The second requires basic multiplication operation of sparse matrices.

8.1 Discrete Laplace-Beltrami Operator

The Laplace-Beltrami operator is fundamental for differential geometry on Riemannian manifolds. Let (M, μ) be a n-D compact Riemannian manifold possibly with boundary ∂M and volume μ defined by

$$d\mu = \sqrt{g}dx^1 dx^2 \dots dx^n, \tag{8.1}$$

where g is the metric of M, and $x^1, x^2, \dots, x^n$ are coordinates in the local chart. The Laplace-Beltrami operator is defined as the divergence of the gradient,

$$\Delta f = \text{div grad } f. \tag{8.2}$$

In a local chart, the gradient of a scalar field f along one direction is given by

$$(\text{grad } f)^i = \partial^i f = g^{ij} \partial_j f, \tag{8.3}$$

and the divergence of a vector field F can be written as

$$\text{div } F = \frac{1}{\sqrt{|g|}} \partial_i (\sqrt{|g|} F^i). \tag{8.4}$$

Therefore, the Laplace-Beltrami operator on manifold M is formulated as

$$\Delta_M = \frac{1}{\sqrt{g}} \sum \frac{\partial}{\partial x^i} \left(g^{ij} \sqrt{g} \frac{\partial}{\partial x^j} \right), \qquad (8.5)$$

where $(g^{ij}) = (g_{ij})^{-1}$, and $g = \det(g_{ij})$.

8.1.1   Cotangent Scheme

On a graph, the Laplacian operator is given by the Laplacian matrix L with entries

$$l_{i,j} = \begin{cases} -1, & \text{if } i \neq j \text{ and } j \in N(x_i) \\ \deg(x_i), & \text{if } i = j, \\ 0, & \text{otherwise} \end{cases} \qquad (8.6)$$

where $\deg(x_i)$ is the degree of node i, and $N(x_i)$ is the neighbor, or directly connected nodes of i.

A normal graph simply measures the connectivity between two adjacent nodes as "1". Weighted graphs measure the connectivity by the weight between two adjacent nodes,

$$l_{i,j} = \begin{cases} -w_{ij}, & \text{if } i \neq j \text{ and } j \in N(x_i) \\ \sum_k w_{ik}, & \text{if } i = j, \\ 0, & \text{otherwise} \end{cases} \qquad (8.7)$$

The Laplacian matrix is the difference of the degree matrix D and the adjacency matrix A of the graph,

$$L = D - A, \qquad (8.8)$$

where D is a diagonal matrix and A is a symmetric matrix. The normalized Laplacian matrix is given by

$$L' = D^{-1/2} L D^{-1/2} = I - D^{-1/2} A D^{-1/2}. \qquad (8.9)$$

On discrete surfaces, a discrete Laplace-Beltrami operator K is given by [Meyer et al. 02]

$$K(x_i) = \frac{1}{2A_i} \sum_{x_j \in N(x_i)} (\cot \alpha_{ij} + \cot \beta_{ij})(x_i - x_j), \qquad (8.10)$$

where A_i is the Voronoi region area of x_i, $N(x_i)$ is the 1-ring neighborhood of x_i, and α_{ij} and β_{ij} are the two angles opposite to the edge (x_i, x_j), as

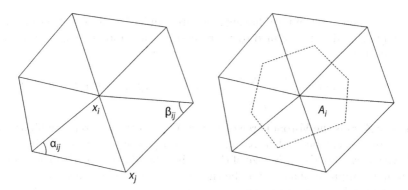

Figure 8.1. Discrete Laplace-Beltrami operator: 1-ring neighbors and angles opposite to an edge (left), and Voronoi region (right).

shown in Fig.8.1. Similarly, the Laplace-Beltrami operator can be implemented as a Laplace-Beltrami matrix L with entries

$$l_{i,j} = \begin{cases} -\frac{\cot \alpha_{ij} + \cot \beta_{ij}}{2A_i}, & \text{if } i \neq j \text{ and } j \in N(x_i) \\ \sum_k \frac{\cot \alpha_{ik} + \cot \beta_{ik}}{2A_i}, & \text{if } i = j, \\ 0, & \text{otherwise} \end{cases} \qquad (8.11)$$

The Voronoi region area can be computed by

$$A_i = \frac{1}{8} \sum_{j \in N(x_i)} (\cot \alpha_{ij} + \cot \beta_{ij}) \|x_i - x_j\|^2. \qquad (8.12)$$

8.1.2 Belkin Scheme

The above discrete Laplace-Beltrami operator is known as the cotangent scheme, which is widely used in computer graphics. However, it has been shown non-convergent for functions other than linear functions [Belkin et al. 08]. Specifically, [Belkin et al. 08] proposed a mesh Laplace operator L_M^h. It takes a function f, and computes

$$L_M^h f(x) = \frac{1}{4\pi h^2} \sum_{t \in M} \frac{\text{Area}(t)}{\#t} \sum_{y \in V(t)} e^{-\frac{\|x-y\|^2}{4h}} (f(y) - f(x)), \qquad (8.13)$$

where h is a parameter, M is a mesh, t is a face, $\text{Area}(t)$ is the area of t, and $\#t$ is the number of vertices of t.

The mesh Laplace operator is expressed with high complexity. The first impression is, for a point x, it involves all faces of the mesh. In fact, considering the exponent $e^{-\frac{\|x-y\|^2}{4h}}$ decreases dramatically when y goes far from

x, only a small set of neighboring faces are required. The size of the affected neighborhood is determined by parameter h. Therefore, we can rewrite this equation to

$$L_M^h f(x) = \frac{1}{4\pi h^2} \sum_{y \in N_h(x)} A(x) e^{-\frac{\|x-y\|^2}{4h}} (f(y) - f(x)), \qquad (8.14)$$

where $A(x)$ is the Voronoi area of x, and $N_h(x)$ is the affected neighborhood of x dependent on h. The parameter h is a positive quantity, which intuitively corresponds to the size of the neighborhood considered at each point. In many applications and for the theoretical analysis in [Belkin et al. 08] h is taken to be independent with x. However, in general, h can be a function of x, which will allow the operator to adapt to the local mesh size. This discrete Laplace-Beltrami operator is related to the Gaussian

$$G_t(x, y) = \frac{1}{4\pi t} e^{-\frac{\|x-y\|^2}{4t}}, \qquad (8.15)$$

which is the heat kernel in Euclidean space. According to Chapter 5, by taking partial derivative with respect to t, we have the Mexican hat wavelet $\psi_t(x, y)$, whose convergence at $t \to 0$ is the Laplace-Beltrami operator.

[Belkin et al. 08] exhibited the convergence of the mesh Laplace operator by the following theorem.

Theorem 8.1 ([Belkin et al. 08]) *Let M be an (ϵ, η)-approximation of a surface S. Put*

$$h(\epsilon, \eta) = \epsilon^{\frac{1}{2.5+\alpha}} + \eta^{\frac{1}{2.5+\alpha}}$$

for an arbitrary fixed positive number $\alpha > 0$. Then for any $f \in C^3(S)$

$$\lim_{\epsilon, \eta \to 0} \sup_{M_{\epsilon, \eta}} \|L_{M_{\epsilon,\eta}}^{h(\epsilon,\eta)} f - \Delta_S f\|_\infty = 0 \qquad (8.16)$$

when the supremum is taken over all (ϵ, η) approximation of S.

For the proof, please refer to [Belkin et al. 08].

8.1.3 Wavelet Scheme

From Chapter 5, we know that the short-time convergence of the Mexican hat wavelet is the Laplace-Beltrami operator

$$\lim_{t \to 0} \psi_t(x, y) = \Delta_M(x, y). \qquad (8.17)$$

The Mexican hat wavelet is defined as negative first derivative of the heat kernel

$$\psi_t(x, y) = -\frac{\partial h_t(x, y)}{\partial t}. \qquad (8.18)$$

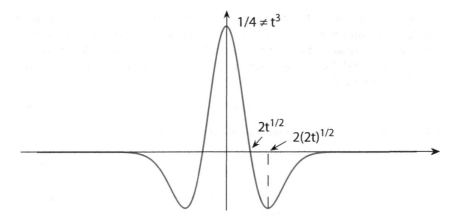

Figure 8.2. Illustration of the approximate Laplace-Beltrami operator along one direction at a small t. Along $|x - y|$, it crosses zero at $|x - y| = 2\sqrt{t}$, and reaches minimum at $|x - y| = 2\sqrt{2t}$.

Also, the heat kernel in 2D Euclidean space has an explicit expression as a Gaussian

$$G_t(x,y) = \frac{1}{4\pi t} e^{-\frac{\|x-y\|^2}{4t}}. \tag{8.19}$$

Then the Mexican hat wavelet in Euclidean space can be written as

$$\psi_t(x,y) = -\frac{\partial G_t(x,y)}{\partial t} = \left(\frac{1}{4\pi t^2} - \frac{\|x-y\|^2}{16\pi t^3}\right) e^{-\frac{\|x-y\|^2}{4t}}. \tag{8.20}$$

By selecting a small t, heat kernel $h_t(x, \cdot)$ is mostly influenced by a small area around x. On 2D manifolds, this is close to a 2D planar area in Euclidean space. Therefore, we obtain an approximation of the Laplace-Beltrami operator

$$\Delta_M(x,y) = \left(\frac{1}{4\pi t^2} - \frac{\|x-y\|^2}{16\pi t^3}\right) e^{-\frac{\|x-y\|^2}{4t}}. \tag{8.21}$$

As shown in Fig. 8.2, the approximate Laplace-Beltrami operator has a similar shape with the Mexican hat wavelet. Its zero-crossing point along $|x - y|$ is at $|x - y| = 2\sqrt{t}$, computed by solving

$$\frac{1}{4\pi t^2} - \frac{\|x-y\|^2}{16\pi t^3} = 0. \tag{8.22}$$

And its minimum point is at $|x - y| = 2\sqrt{2t}$, computed by solving

$$\frac{\partial \Delta_M(x,y)}{\partial |x-y|} = 0. \tag{8.23}$$

Hence, for a given small t, we will need a supporting area with radius at least $2\sqrt{2t}$ for computing the Laplace-Beltrami operator. The discrete operator is computed by sampling by vertices in this supporting area. Assuming the normalized mesh has average edge length as one unit, we get a lower bound of t by setting $2\sqrt{2t} = 1$:

$$t_{min} = \frac{1}{8}. \tag{8.24}$$

An appropriate value of small t is obtained by setting $2\sqrt{t} = 1$

$$t = \frac{1}{4}. \tag{8.25}$$

The above is a continuous formulation in spatial domain. For a discrete mesh, samples need to be made at each vertex. Consider a vertex $x \in M$; for any $y \neq x$ with $|x - y| < 2\sqrt{2t}$, we define the discrete Laplace-Beltrami operator as

$$L_M(x, y) = A(y) \left(\frac{1}{4\pi t^2} - \frac{\|x - y\|^2}{16\pi t^3} \right) e^{-\frac{\|x-y\|^2}{4t}}. \tag{8.26}$$

The Mexican hat wavelet has zero-mean property

$$\int_M \psi_t(x, y) d\mu(y) = 0. \tag{8.27}$$

To keep this property, we let

$$L_M(x, x) = -\sum_y L_M(x, y). \tag{8.28}$$

Leaving area weight out of account, this formulation is slightly different with the Belkin scheme by one term

$$\frac{\|x - y\|^2}{16\pi t^3} e^{-\frac{\|x-y\|^2}{4t}}.$$

When $t \to 0$, this term is more significant than the other one

$$\frac{1}{4\pi t^2} e^{-\frac{\|x-y\|^2}{4t}}$$

for $x \neq y$.

8.1.4   Point Cloud

Point cloud is also a widely-seen format for 3D geometry data. By the Belkin scheme and the wavelet scheme, we can compute the Laplace-Beltrami operator on point clouds without triangulation. Assuming there

---

Algorithm 2: Discrete Laplace-Beltrami matrix.

Input: data M

Output: discrete Laplace-Beltrami matrix L

Initialize L as a zero matrix;

for *each* $x \in M$ **do**

　Find a set of neighbors $N(x)$;

　for *each* $y \in N(x)$ **do**

　　Compute $L(x, y)$ by a discretization scheme;

　end

end

---

is no collision problem, all we need is a set of nearest neighbors for each point. This can be achieved by the approximate nearest neighbor (ANN) algorithm [Arya et al. 98].

A general algorithm for computing discrete Laplace-Beltrami matrix is documented in Algorithm 2. It can be applied to both meshes and point clouds. For meshes, the neighborhood $N(x)$ is obtained directly by mesh connectives subject to certain conditions. For point clouds, it is obtained by the ANN package. It is worth mentioning that the computed matrix L is heavily sparse, since only a few entries are non-zero for each data point. The sparsity is convenient for fast computation.

8.2　Generalized Eigenvalue Problem

8.2.1　Implementation Method

The computation of heat kernels and wavelets on manifolds often need to solve eigenvalues and eigenfunctions of the Laplace-Beltrami operator. Given a discrete Laplace-Beltrami operator L, we have

$$L\phi_k(x) = -\lambda\phi_k(x), \qquad (8.29)$$

where λ_k and $\phi_k(x)$ are the k-th eigenvalue and eigenfunction of the operator. The matrix L is supposed to be self-adjoint, i.e. symmetric, in order to have complete orthogonal eigenfunctions. However, from the above section, we know discrete Laplace-Beltrami operators are usually not symmetric, especially for meshes with non-uniform sampling.

A discrete Laplace-Beltrami matrix L can be decomposed to the product of a diagonal matrix A and a symmetric matrix W

$$L = A^{-1}W. \qquad (8.30)$$

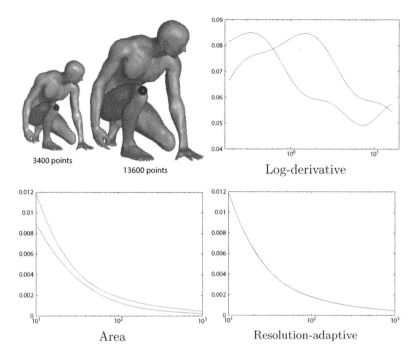

Figure 8.3. Two corresponding heat kernel signature curves are computed on scaled, resolution-varying shapes, under different scaling schemes. The log-derivative plot for the Fourier scheme has non-constant boundaries. The area scheme has meaningful time selection, but also a time-shift and a scale change. The resolution-adaptive area scheme solves this problem.

The symmetric matrix W is highly sparse, since the operator at a vertex only affects a small neighborhood. The diagonal matrix A is related to Voronoi areas. If a mesh has uniform sampling, all vertices will have the same Voronoi areas, and therefore, L will be symmetric. For general cases, L is non-symmetric. By applying the decomposition, the Eq.(8.29) turns to a generalized eigenvalue problem

$$W\phi_k(x) = -\lambda A\phi_k(x). \tag{8.31}$$

The generalized eigenvalue problem of sparse matrix can be solved by the Matlab method "eigs"

$$[\text{eigenfunctions eigenvalues}] = \text{eigs}(W, A, K).$$

This method returns K eigenfunctions and eigenvalues for the Eq.(8.31).

The eigen-system has a scale problem derived from the area matrix A in the Laplace-Beltrami operator. Taking the heat kernel as an example, there are some scaling schemes in existing work, such as eigenvalue [Sun et al. 09], Fourier transform [Bronstein and Kokkinos 10], and area [Vaxman et al. 10]. The eigenvalue scheme scales eigenvalues by a reciprocal of the first non-trivial eigenvalue. It does not indicate how to select values of t. The Fourier scheme first processes heat kernels by logarithm and derivative, which transfers the scale problem to a shift in logarithmic time. Then, it utilizes Fourier transform to eliminate the shift, leading to scale-invariant heat kernels. However, it has boundary problems in Fourier transform. Heat kernels have Gaussian decay that drops exponentially in both spatial and temporal domains [Boggess and Raich 10], whose boundaries are non-constant. As shown in Fig. 8.3, we compute corresponding HKS values on scaled, resolution-varying shapes. The log-derivative plot for the Fourier scheme shows that the two HKS curves have different boundaries besides a time-shift. The area scheme eliminates the scale factor by enforcing the average vertex area to be unit one. It also makes the selection of t meaningful, such that $t = 1$ results in an average influence region of about 1-ring. This scaling is naturally invariant to scale changes. However, when the mesh resolution changes, the unit area covers regions with different sizes, causing a shift and a scale for heat kernels, as shown by the area-scheme plot in Fig. 8.3.

To combat the aforementioned problems, we design a resolution-adaptive area scheme, which is meaningful in time selection and invariant to resolution changes. Assume a surface is resembled by α times points. Since the area scheme changes the area unit to be one, the total area has the same ratio plugged into the generalized eigen-decomposition. It shifts logarithm of t backward by $\log \alpha$, and scales heat kernels by $1/\alpha$. There, we compensate t and heat kernels by these amounts, leading to a result shown by the resolution-adaptive plot in Fig. 8.3. This scheme is adaptive to scale and resolution changes.

8.2.2 Evaluation

A small number of eigenvalues and eigenfunctions $\{\lambda_k, \phi_k(x)\}_{k=0}^{K}$ can be used to compute heat kernels and some diffusion-related quantities. The heat kernel can be approximated by

$$h_t(x, y) = \sum_{k=0}^{K} e^{-\lambda_k t} \phi_k(x) \phi_k(y). \tag{8.32}$$

The Mexican hat wavelet from Chapter 5 can be approximated by

$$\psi_t(x, y) = \sum_{k=0}^{K} \lambda_k e^{-\lambda_k t} \phi_k(x) \phi_k(y). \tag{8.33}$$

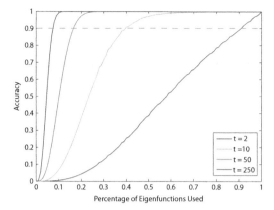

Figure 8.4. Accuracy of Mexican hat wavelet $\psi_t(x,x)$ on a 400-point 1D manifold, computed with different percentages of eigenfunctions. A high accuracy can be achieved by increasing the number of eigenfunctions used. High frequencies (small t) require more eigenfunctions to maintain accuracy. The determination of the number of eigenfunctions depends on t and the data size.

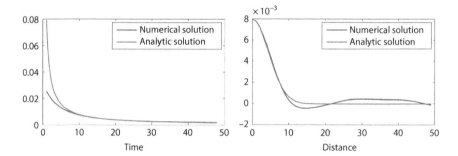

Figure 8.5. Numerical and analytic heat kernels on a triangular mesh of a 2D plane: $h_t(x,x)$ with varying t (left); $h_t(x,y)$ with $t = 10$ and varying distance $|x - y|$ (right).

We evaluate the accuracy through computing the Mexican hat wavelet. For meshed surfaces, we approximate wavelets and their transforms by a relatively small number of eigenfunctions.

As empirical suggestions, for $3\% \sim 5\%$ eigenfunctions, $t \geq 8$ would be a reasonable setting. For example, in Fig. 8.5, we compare the numerical solution (using 3% eigenfunctions) and analytic solution

$$\frac{1}{4\pi t} e^{-\frac{\|x-y\|^2}{4t}}$$

Algorithm 3: Wavelet transform by convolution.

Input: eigen-system $\{\lambda_k, \phi_k(x)\}$, function $f(x)$, time t
Output: wavelet transform $\mathcal{W}_f(x, t)$

for *each* $x \in M$ **do**
    for *each* $y \in M$ **do**
        if $\psi_t(x, y)$ *not computed* **then**
            Initialize $sum = 0$;
            for $k = 0 : K$ **do**
                $sum\ +\ = \lambda_k e^{-\lambda_k t} \phi_k(x) \phi_k(y)$;
            end
            $\psi_t(x, y) = sum$;
        end
    end
end
for *each* $x \in M$ **do**
    Initialize $sum = 0$;
    for *each* $y \in M$ **do**
        $sum\ +\ = \psi_t(x, y) f(y)$;
    end
    $\mathcal{W}_f(x, t) = sum$;
end

---

on a triangular mesh of a 2D plane. Using the area scheme, t is meaningful and equivalent to the one in the analytic solutions. For very small t, the numerical and analytic solutions have large differences. Furthermore, the numerical solutions start to oscillate around the analytic solution when the distance $|x - y|$ is getting larger.

This scheme has been widely used for computing heat kernels on 3D shapes [Sun et al. 09, Ovsjanikov et al. 10, Bronstein and Kokkinos 10, Bronstein et al. 11, Hou and Qin 12a]. It can conveniently access all frequencies corresponding to t with the same cost. Through the above evaluation, we see that it reaches high accuracy for low frequencies corresponding to large t. Due to the small number of used eigenfunctions, it is limited to high frequencies corresponding to small t.

Algorithm 4: Wavelet transform by Fourier transform.

Input: eigen-system $\{\lambda_k, \phi_k(x)\}$, function $f(x)$, time t
Output: wavelet transform $\mathcal{W}_f(x, t)$

for $k = 0 : K$ **do**
    Initialize $sum = 0$;
    for *each* $x \in M$ **do**
        $sum + = \phi_t(x)f(x)$;
    end
    $\widehat{f}(k) = sum$;
end
for *each* $x \in M$ **do**
    Initialize $sum = 0$;
    for $k = 0 : K$ **do**
        $sum + = \lambda_k e^{-\lambda_k t} \widehat{f}(k)\phi_k(x)$;
    end
    $\mathcal{W}_f(x, t) = sum$;
end

8.2.3   Wavelet Transform Implementation

As introduced in Chapter 5, the continuous wavelet transform (CWT) of the Mexican hat wavelet is given by

$$\psi_t(x, y) = \sum_{k=0}^{\infty} \lambda_k e^{-\lambda_k t} \phi_k(x)\phi_k(y).$$

There are two methods for computing the wavelet transform: convolution and Fourier transform. For a given function $f(x) \in L^2(M)$, the wavelet transform computed by convolution refers to the inner product

$$\mathcal{W}_f(x, t) = \langle \psi_t(x, y), f(y) \rangle. \tag{8.34}$$

The wavelet transform computed by Fourier transform refers to

$$\mathcal{W}_f(x, t) = \sum_{k=0}^{\infty} \lambda_k e^{-\lambda_k t} \widehat{f}(k)\phi_k(x), \tag{8.35}$$

where $\widehat{f}(k)$ is the Fourier transform of $f(x)$.

Assume the number of vertices is V, and the number of used eigenfunctions is K. The convolution method for computing CWT is documented in Algorithm 3. The time complexity to compute all wavelets $\psi_t(x, y)$ at

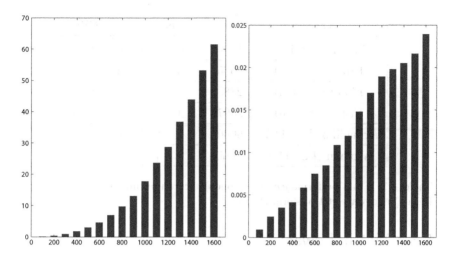

Figure 8.6. Time performance of computing wavelet transform by convolution (left) and Fourier transform (right), with 10% eigenfunctions used. The horizontal axis denotes data size. The Fourier method significantly improves the time performance.

a t is then $O(KV^2)$, and so is the convolution-based wavelet transform in Eq. (8.34). It is extremely inconvenient for practical use, since V is usually at the order of tens, maybe even hundreds of thousands. Similar to the case in signal processing, the proposed Fourier method in Eq.(8.35) with a spectral expression significantly improves the computation, documented in Algorithm 4. It has $O(KV)$ time complexity, with the same accuracy of the convolution. Fig. 8.6 shows the time performance of a wavelet transform computed by convolution (left) and Fourier transform (right), with 10% eigenfunctions used. The computation time does not include the time for computing Laplace-Beltrami eigenfunctions. The horizontal axis denotes data size. The performance of the spectral expression is approximately linear to the data size, which significantly improves the time performance. The spectral expression also enables an efficient method to compute the heat transform and solve the heat equation.

Table 8.1 documents time performance for computing eigen-decomposition (Eigen) and some levels of continuous wavelet transform (CWT) on some 3D models that will appear in Chapter 10 and Chapter 11. Here, we use $K = 300$ for all experiments, running on a laptop with Core2 Duo CPU 2.53GHz and 4GB RAM. The computation bottleneck lies in solving the generalized eigenvalue problem. Upon solving that, the efficiency of applying wavelet transforms can be significantly improved.

Data	V	Level	Eigen	CWT
Rocker (Fig. 10.5)	10044	5	30.16	0.12
Eros (Fig. 10.6)	25651	3	69.62	0.26
Lion (Fig. 10.6)	23889	4	64.38	0.29
Fish (Fig. 10.6)	24975	4	67.59	0.34
Armadillo (Fig. 11.7)	50000	5	139.93	1.09
Woman (Fig. 11.8	9971	3	28.36	0.11

Table 8.1. Time performance (in seconds) for computing eigen-decomposition (Eigen) and wavelet transform (WT).

8.3   Matrix Power

Another method to implement diffusion is based on a local operator and its semigroup identity, for example the implementation in Chapter 4. For small t, the heat kernel can be approximated by its Gaussian operator. One can use this short-time approximation to compute long-time heat kernels and heat operators. For example, methods in [Lee et al. 05, Zaharescu et al. 09] adopted a Gaussian kernel with Euclidean metric at small t, and in [Hou and Qin 10, Zou et al. 08], a Gaussian kernel with geodesic metric was used to approximate the short-time heat kernel. In [Hou and Qin 13], this scheme is extended to any diffusion-type local operators.

When coming to large scales, this approximation is very limited at accuracy and computational speed. To address this problem, the semigroup identity is often adopted to compute large-scale diffusion by small-scale diffusion. By the semigroup identity, we have $H_{2t} = H_t H_t$. This operation can applied iteratively, leading to

$$H_{nt} = \underbrace{H_t H_t \cdots H_t}_{n} = H_t^n. \tag{8.36}$$

8.3.1   Sparse Matrix Multiplication

More practically, this multiplication is formulated as a dyadic power series of some small t like our work in [Hou and Qin 13]

$$H_{2^j t} = H_t^{2^j}. \tag{8.37}$$

This reduces the computation to one operation: matrix square, i.e. multiplication of two identical matrices. As it always starts from some small t for accuracy and efficiency, the matrix H_t is highly sparse, where only non-zero elements are involved in the computation. To save computation,

---

Algorithm 5: Admissible diffusion wavelets.

Input: data M, levels J, parameter ϵ

Output: scaling matrix $\{\Phi_j\}_{j=0}^{J}$, wavelet matrix $\{\Psi_j\}_{j=1}^{J}$

Compute operator T;

$\Phi_0 = I$, $\Phi_1 = [\Phi_0 T]_r$;

$\Psi_1 = \Phi_0 - \Phi_1$;

for $j = 2 : J$ **do**

$\quad \Phi_j = [\Phi_{j-1}\Phi_{j-1}]_{r,\epsilon}$;

$\quad \Psi_j = \Phi_{j-1} - \Phi_j$;

end

---

near-zero elements can be filtered by some threshold. For very large scales, heat operator matrices are not sparse. Then, the matrix multiplication is better to be computed by software packages like multi-thread computing in MATLAB and parallel computing by GPU.

The eigen-system and matrix power are two fundamental solutions for implementing heat kernels and diffusion. The matrix-power solution is suitable for small-time diffusion, where it has high accuracy and fast computation. For large values of t, it is costly to compute powers of the heat operator. On the contrary, the eigen-system solution is more accurate for long-time heat kernels, since large t corresponds to large scales that do not need many eigenfunctions. But it has limitation in short time which requires high-frequency eigenvalues and eigenfunctions. As the third scheme proposed in [Vaxman et al. 10], Vaxman et al. showed that heat kernels at any time can be properly approximated by a lower resolution version of the original surface. As a compromise, a multiresolution approach was proposed by mixing these two methods, for a fast, accurate solution that adapts to a large range of t. This is a compound of the schemes of eigen-system and matrix power.

8.3.2   Implementation of Admissible Diffusion Wavelets

Diffusion wavelets introduced in Chapter 4 can be implemented by matrix multiplication. Time performance is considered as an important criterion for the algorithm design. Algorithm 5 documents the construction of scaling functions and wavelets of the diffusion wavelets.

For a given data (mesh or point cloud) M, the method first computes the local operator T. Then, it initializes the first level of scaling functions Φ_1 and wavelets Ψ_1, stored as sparse matrices. The following levels of scaling functions are computed by sparse matrix multiplication of previous levels,

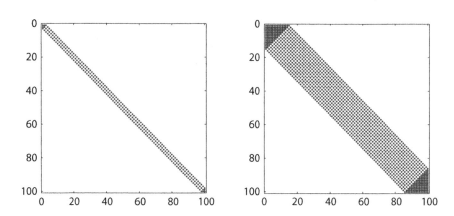

Figure 8.7. Sparse matrices Φ_1 and Φ_4 from 1D manifold with 100 points.

subject to normalization and a threshold ϵ. Finally, the differences of two adjacent scaling functions form the wavelets. The computation of diffusion wavelets is mainly on non-zero entries of sparse matrices.

In [Hou and Qin 13], we used MATLAB to compute the operations of sparse matrices, whose complexity is linear to the number of non-zero elements. Assume the number of vertices is V, and for all $x \in M$, the operator $T(x, \cdot)$ has at most N neighboring entries, where N is a constant. The time complexity for computing j levels of diffusion wavelets is $O(2^{2j}V)$. The amount of computation will be huge for even moderate values of j. Since we are particularly interested in high-frequency processing of spectral analysis, a small value of $j \leq 8$ is sufficient for the applications. In this case, Φ_j and Ψ_j continue to be sparse matrices. To ensure the numerical sparsity, ignorable elements are eliminated by a threshold $\epsilon = 10^{-6}$ before the row-based normalization.

Fig. 8.7 visually depicts the structure of the sparse matrices Φ_1 and Φ_4 from 1D manifold with 100 points. The matrices are sparser for data with more points. For other applications that need to access lower frequencies, one could down-sample the scaling transforms to reduce computation.

8.3.3 Time Performance

In [Hou and Qin 13], we examined the time efficiency of the admissible diffusion wavelets in the applications, with performance being documented in Table 8.2. The prototype software was developed on a laptop with Core2 Duo CPU 2.53GHz and 4GB RAM. In each experiment, we recorded the computation time of the wavelet system $\{\Phi_j, \Psi_j\}$ denoted as "ADW", and

Data	# V	J	ADW	Processing
Gargoyle (Fig. 10.2)	25011	7	28.63	0.11
Buddha (Fig. 10.2)	25003	7	30.15	0.12
Hand (Fig. 10.2)	10000	7	14.66	0.05
Sphere (Fig. 10.3)	10242	8	30.34	0.05
Skull (Fig. 10.4)	20002	7	19.39	0.11
Dinosaur (Fig. 11.1)	14053	8	35.03	0.03
Face (Fig. 11.3)	10156	5	1.67	0.47
Pegaso (Fig. 11.4)	30000	4	7.12	1.25
Kitten (Fig. 11.4)	28969	4	4.09	1.18
Horse (Fig. 11.5)	10296	4	1.70	0.39
Cow (Fig. 11.5)	16612	4	3.11	0.62
Armadillo (Fig. 11.6)	50000	6	30.61	5.52

Table 8.2. Time performance (in seconds) of the admissible diffusion wavelets.

the processing time for its application. The computation of the ADW, depending on the number of points and the number of levels, is very efficient, so they can afford rapid spectral analysis. The processing varies for different applications in Chapter 10 and Chapter 11, with time recorded in the column "Processing".

9

Shape Representation

3D shapes are often represented by polygonal meshes or point clouds. These representations are convenient for rendering, but they are not capable of many applications in computer graphics and computer vision. For example, compare two shapes undergoing deformation, which cannot transform from one to another by rotation and translation.

This chapter presents some recent results in the research of shape representation, including heat kernel signature [Sun et al. 09], wave kernel signature [Aubry et al. 11], and wavelet signature [Hou and Qin 12a]. These representations are intrinsic to shape geometry, and therefore, invariant to isometric deformation.

9.1 Related Work

There is a rich literature on shape representation. Early research work is concerned more with rigid shapes. A great reference is the 3D shape retrieval project of Princeton University [1]. Some representative shape descriptors studied in this project include shape distributions, reflective symmetry descriptors, spherical harmonics, and skeletal graphs [Kazhdan 04].

Shape distribution represents the shape of a 3D model as a probability distribution sampled from a shape function measuring geometric properties. The work in [Osada et al. 01] models 3D shapes by generating a histogram of distances between pairs of points on the shape. Reflective symmetry descriptors utilize symmetries of 3D shapes. In [Kazhdan et al. 02], a descriptor was defined as a representation of the symmetry distances for all planes through the model's center of mass. Representations based on spherical harmonics project a shape to a complete basis of sphere Laplace operator [Kazhdan et al. 03]. They are invariant to rotation. Skeletal graph is a high-level representation that models a shape by a graph of its skeleton.

---

[1] http://gfx.cs.princeton.edu/proj/shape/

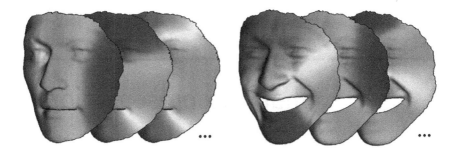

Figure 9.1. Some Laplace-Beltrami eigenfunctions of two partial shapes undergoing deformation and boundary changes. The two shapes share most common part of the object. However, the orders of their eigenfunctions vary due to deformation and boundary changes. See color insert.

When deformation is considered, it is more challenging to define shape representation. We need new tools to compute descriptors that are invariant to deformation. One of such tools is the eigen-system of the Laplace-Beltrami operator. As introduced before, the Laplace-Beltrami eigenfunctions (i.e. the manifold harmonics) compose an orthogonal and complete basis of the function space. In [Rustamov 07], a shape descriptor was defined as the vector

$$\left(\frac{1}{\sqrt{\lambda_1}}\phi_1(x), \frac{1}{\sqrt{\lambda_1}}\phi_1(x), \dots\right). \tag{9.1}$$

This was called the global point signature (GPS). A series of Laplace-Beltrami eigenvalues is the spectrum of the shape, which is characteristic. In [Reuter et al. 06], it was called the "shape-DNA". The eigenvalues of eigenfunctions are very good shape representations, as they are invariant to isometric deformation. In this book, we categorize them as global descriptors, because they are determined by the entire shape.

Fig. 9.1 shows some Laplace-Beltrami eigenfunctions of two partial shapes undergoing deformation and small boundary changes. The two shapes share most common part of the object. However, the orders of their eigenfunctions vary due to deformation and boundary changes. It indicates that the Laplace-Beltrami eigen-system is capable as representation of partial shapes with boundary changes. In fact, the order of Laplace-Beltrami eigenfunctions is unstable for shapes with repeated or near eigenvalues. A small perturbation can cause order of two eigenfunctions with near eigenvalues to flip.

Accordingly, shape descriptors that are determined by a local area of the shape are referred as local descriptors. One such example is the heat

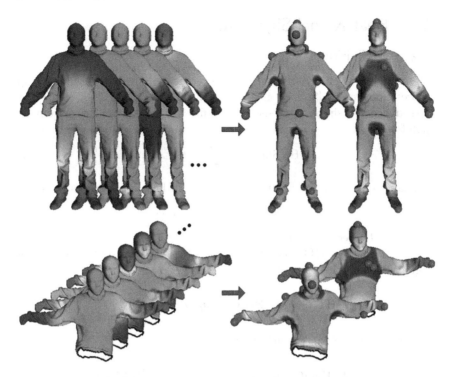

Figure 9.2. Shape representation methods. Two shapes that are partially over-lapped with deformation have different Laplace-Beltrami eigenfunctions (left) but similar heat kernel signatures (right). See color insert.

kernel signature [Sun et al. 09]. As introduced before, heat kernels are lo-cally supported by neighboring areas. Heat kernels are multiscale, which naturally connect local and global geometry. Moreover, the heat kernel signature captures all geometric information up to isometry. That is, two shapes with isometric deformation have exactly the same heat kernel sig-natures, and vice versa. This property was called intrinsic in [Sun et al. 09]. As a supreme shape representation, heat kernel signature has been adopted in many applications including shape matching [Ovsjanikov et al. 10], reg-istration [Hou and Qin 12b], and retrieval [Bronstein et al. 11, Ovsjanikov et al. 09].

Fig. 9.2 shows Laplace-Beltrami eigenfunctions and heat kernel sig-natures as shape representation methods for partially overlapped shapes. The two shapes have dramatically different eigenfunctions but very simi-lar heat kernel signatures. This is because Laplace-Beltrami eigenfunctions are global, which are easily affected by boundary changes, and heat kernel signatures are local, which are stable within local areas.

9.2   Heat Kernel Signature

9.2.1   Definition

Signatures that characterize the shape geometry have been used for shape representation. In [Sun et al. 09], the heat kernel signature (HKS) was introduced for shape analysis. It is defined as the heat kernel from one point to itself

$$\text{HKS}(x,t) = \sum_{k=0}^{\infty} e^{-\lambda_k t} \phi_k^2(x). \tag{9.2}$$

A shape representation is given by a vector of heat kernel signatures

$$[\text{HKS}(x,t_0),\ \text{HKS}(x,t_1),\ \cdots]^T$$

sampled at a sequence of time.

The heat trace (HT) on manifold M is the integration of all heat kernel signatures over the domain

$$\text{HT}(t) = \int_M \text{HKS}(x,t)d\mu(x) = \sum_{k=0}^{\infty} e^{-\lambda_k t}. \tag{9.3}$$

Proof: The proof is straightforward:

$$\int_M \text{HKS}(x,t)d\mu(x) = \int_M \sum_{k=0}^{\infty} e^{-\lambda_k t}\phi_k^2(x)d\mu(x)$$

$$= \sum_{k=0}^{\infty} e^{-\lambda_k t} \int_M \phi_k^2(x)d\mu(x) \qquad \square$$

$$= \sum_{k=0}^{\infty} e^{-\lambda_k t}$$

In [Sun et al. 09], the heat trace is used to scale shape representation by heat kernel signature:

$$\left[\frac{\text{HKS}(x,t_0)}{\text{HT}(t_0)}, \frac{\text{HKS}(x,t_1)}{\text{HT}(t_1)}, \cdots\right]^T.$$

This is to balance heat kernel signatures at different scales, since the heat kernel signature drops exponentially along time t, as shown in Fig. 9.3. However, this method is not applicable for partial shapes with changing boundaries. The heat trace is associated with a manifold M and a time t. In [Hou and Qin 12b], we used another scheme for scaling heat kernel signatures. Consider the heat kernel in $\mathbb{R}^2$

$$G_t(x,y) = \frac{1}{4\pi t} e^{-\frac{|x-y|^2}{4t}},$$

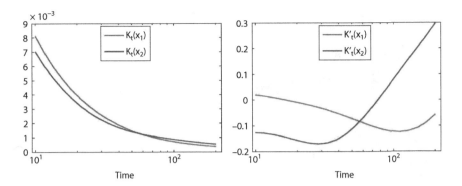

Figure 9.3. Semi-log plots of the heat kernel signature (left) and the rescaled heat kernel signature (right).

which is a Gaussian. The heat kernel signature is then reduced to

$$g(t) = \frac{1}{4\pi t}.$$

We adopted it to rescale the shape representation to

$$\left[\log\left(\frac{\text{HKS}(x, t_0))}{g(t_0)}\right), \log\left(\frac{\text{HKS}(x, t_1)}{g(t_1)}\right), \cdots\right]^T.$$

It approximately reflects how a surface is curved from plane with respect to t. Consequently, this normalization characterizes how the surface is bent from the plane, from diffusion's perspective. Positive values imply slower heat-diffusion than a plane, while negative values imply faster diffusion. Fig. 9.3 shows the heat kernel signature before and after rescaling, where the HKS is denoted as

$$K_t(x) : \text{HKS}(x, t),$$

and the rescaled HKS is denoted as

$$K_t'(x) : \log\left(\frac{\text{HKS}(x, t))}{g(t)}\right).$$

Fig. 9.4 shows three normalized HKS curves. This flexible normalization balances components at different values of t, and helps discriminate features. In HKS descriptors, we logarithmically sample t in the range $[5, 640]$.

9.2.2 Properties

The heat kernel signature has analytical values at short-time and long-time convergences. From Chapter 3, we have the short-time convergence of heat

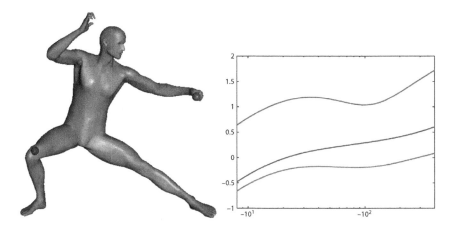

Figure 9.4. Three normalized heat kernel signature curves. The normalization balances components at different values of t. Positive values imply slower heat-diffusion than a plane, while negative values imply faster diffusion.

kernel given by

$$\lim_{t \to 0} h_t(x, y) = \delta(x, y). \tag{9.4}$$

Consequently, the heat kernel signature has

$$\lim_{t \to 0} \text{HKS}_t(x) = \delta(x, x), \tag{9.5}$$

where $\delta(x, x)$ is a Dirac kernel. On discrete surfaces, we have

$$\delta(x, x) = \delta(0) = 1.$$

It can be interpreted in another way

$$\lim_{t \to 0} \text{HKS}_t(x) = \sum_{k=0}^{\infty} \phi_k^2(x), \tag{9.6}$$

which also equals to 1 on discrete spaces.

Also from Chapter 3, we have the long-time convergence of the heat kernel

$$\lim_{t \to \infty} h_t(x, y) = \frac{1}{\mu(M)}, \tag{9.7}$$

where $\mu(M)$ is the Riemannian volume of manifold M, which is a constant. It is also called the stationary state of heat kernels. Fig. 9.5 shows the time behaviors of two corresponding heat kernels on different partial shapes. At the beginning, the heat kernels are close but very small. Then they

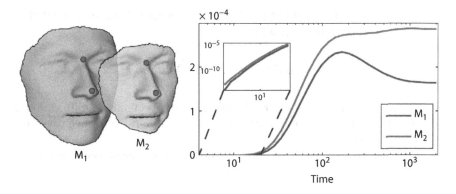

Figure 9.5. The time behaviors of two corresponding heat kernels on partial shapes M_1 (10k vertices) and M_2 (5k vertices). In the semi-log plot, the heat kernels are close but very small at the beginning, with a log-log plot shown nearby. Then they gradually differ due to boundary changes, and finally converge to different stationary states.

gradually differ due to boundary changes, and finally converge to different stationary states. Therefore, the long-time convergence of the heat kernel signature has

$$\lim_{t \to \infty} \text{HKS}_t(x) = \frac{1}{\mu(M)}. \tag{9.8}$$

It is worth mentioning that the first eigenfunction is constant everywhere on the manifold

$$\phi_0(x) = \frac{1}{\sqrt{\mu(M)}}. \tag{9.9}$$

According to the Informative Theorem in [Sun et al. 09], under mild assumptions, heat kernel signature $\{\text{HKS}_t(x)\}_{t>0}$ keeps all of the information of heat kernel $\{h_t(x, \cdot)\}_{t>0}$.

Theorem 9.1 (Informative Theorem [Sun et al. 09]) *If the eigenvalues of the Laplace-Beltrami operators of two compact manifolds M and N are not repeated, and T is a homeomorphism from M to N, then T is isometric if and only if $h_t^M(x, x) = h_t^N(T(x), T(x))$ for any $x \in M$ and any $t > 0$.*

We refer readers to [Sun et al. 09] for a detailed proof of this theorem. Basically, it says the geometry of a shape is characterized up to isometry by its heat kernel signature. If two shapes have same heat kernel signatures at all t, they must be isometry.

As pointed out in [Sun et al. 09], the heat kernel signature is related to the Gaussian curvature $\kappa_G(x)$ by

$$\text{HKS}(x,t) = (4\pi t)^{d/2} \sum_{i=0}^{\infty} a_i t^i, \tag{9.10}$$

where $a_0 = 1$ and $a_1 = \frac{1}{6}\kappa_G(x)$. It reveals that heat tends to diffuse slower at points with positive curvature, and faster at points with negative curvature [Sun et al. 09].

In [Rustamov 07], Rustamov introduced the global point signature (GPS), given by

$$\text{GPS}(x) = \left(\frac{\phi_1(x)}{\sqrt{\lambda_1}}, \frac{\phi_2(x)}{\sqrt{\lambda_2}}, \cdots \right). \tag{9.11}$$

It is a global shape representation, which is determined by the Laplace-Beltrami eigen-system, and hence, the global geometry. As compared with the global point signature, the heat kernel signature can be viewed as a weighted sum of the squares of eigenfunctions [Sun et al. 09].

9.3   Wave Kernel Signature

9.3.1   Definition

In [Aubry et al. 11], the wave kernel signature was introduced by Aubry et al., based on quantum mechanics. The evolution of a quantum particle on manifolds is governed by its wave function $\psi(x,t)$, which is a solution of the Schrödinger equation

$$\frac{\partial \psi}{\partial t}(x,t) = i\Delta_M \psi(x,t), \tag{9.12}$$

where i is the imaginary unit, and Δ_M is the Laplace-Beltrami operator of manifold M. This partial differential equation is similar to the heat equation in terms of formulation, but they are drastically different in induced dynamics [Aubry et al. 11].

Let f_E^2 be an energy probability distribution with expectation value E. Assuming the Laplace-Beltrami operator has no repeated eigenvalues, the wave function of the particle is given by

$$\psi_E(x,t) = \sum_{k=0}^{\infty} e^{i\lambda_k t} \phi_k(x) f_E(\lambda_k), \tag{9.13}$$

where λ_k and $\phi_k(x)$ are the k-th eigenvalue and eigenfunction of the Laplace-Beltrami operator. The probability for the particle at point x is measured by $|\psi_E(x,t)|^2$. The wave kernel signature is then defined as the accumulated probability

$$\mathrm{WKS}(E,x) = \lim_{T\to\infty} \frac{1}{T} |\psi_E(x,t)|^2. \tag{9.14}$$

Considering the orthogonality of the exponent term, [Aubry et al. 11] further simplified the definition to

$$\mathrm{WKS}(E,x) = \sum_{k=0}^{\infty} \phi_k^2(x) f_E^2(\lambda_k). \tag{9.15}$$

Based on the definition above, the problem of shape signature turns to choosing the appropriate distributions f_E^2 and distance between wave kernels. In [Aubry et al. 11], the authors chose a real valued function in the logarithmic energy scale $e = \log(E)$, and rewrite the wave kernel signature as

$$\mathrm{WKS}(x,e) = C_e \sum_k \phi_k^2(x) e^{\frac{-(e-\log \lambda_k)^2}{2\sigma^2}}, \tag{9.16}$$

with

$$C_e = \left(\sum_k e^{\frac{-(e-\log \lambda_k)^2}{2\sigma^2}}\right)^{-1}. \tag{9.17}$$

The distance between two wave kernel signatures is defined as

$$d_{\mathrm{WKS}}(x,y) = \int_{e_{min}}^{e_{max}} \left| \frac{\mathrm{WKS}(x,e) - \mathrm{WKS}(y,e)}{\mathrm{WKS}(x,e) + \mathrm{WKS}(y,e)} \right| de, \tag{9.18}$$

where e_{min} and e_{max} correspond to the largest and smallest considered energy scales. [Aubry et al. 11] stated that the bigger e_{max} the more local information included, and suggested

$$e_{min} = \log(\lambda_1) + 2\sigma$$

and

$$e_{max} = \log(\lambda_K) - 2\sigma.$$

9.3.2 Properties

Some properties of the wave kernel signature are documented in [Aubry et al. 11]. For example, the wave kernel signature is invariant to isometric deformation. This is directly inherited from the Laplace-Beltrami operator. And the wave kernel signature is informative, which characterizes shape geometry up to isometry. This is a similar property to the heat kernel signature.

A more interesting property is the natural notion of scale. [Aubry et al. 11] stated that the wave kernel signature is a function of energy levels that are directly related scales. Large energy levels correspond to highly oscillatory particles, which reflect high wave frequencies mostly affected by local geometry. Small energy levels correspond to low-frequency particles mostly affected by global geometry. This is because the energy level e corresponds to the eigenvalues $\{\lambda_k\}$ of the Laplace-Beltrami operator. As we know, the eigenvalue λ_k corresponds to frequency in Fourier transform. Small eigenvalues correspond to low frequencies and large scales, while large eigenvalues correspond to high frequencies and small scales.

The heat kernel is expanded on the full sequence of eigenfunctions. It uses the diffusion time t to build scales. As t increases, diffusion proceeds to larger scales in spatial domain. On the contrary, the wave kernel signature uses energy level e that corresponds to the Laplace-Beltrami eigenvalue for scales. The heat kernel is multiscale. That is, for a certain time t, heat kernel $h_t(x, \cdot)$ is only influenced by a neighbor area centered at x, i.e. the support area. It drops dramatically outside the support area. This locality property allows the heat kernel signature to capture local geometry. However, the wave function $\psi_E(x, t)$, like the eigenfunctions of the Laplace-Beltrami operator, is a global function over the entire domain. Therefore, the wave kernel signature is a representation for global geometry.

9.4 Wavelet Signature

9.4.1 Definition

As shown in Chapter 5, the Mexican hat wavelet characterizes the shape geometry up to isometry. It is easy to induce a shape representation by considering the wavelet kernel from one point to itself, just like the heat kernel signature. To obtain a more concise formulation, we define the wavelet signature

$$\text{WS}(x, t) : M \times \mathbb{R}^+ \to \mathbb{R}$$

as

$$\text{WS}(x, t) = \psi_t(x, x) = \sum_{k=0}^{\infty} \lambda_k e^{-\lambda_k t} \phi_k^2(x), \qquad (9.19)$$

where

$$\psi_t(x, y) = \sum_{k=0}^{\infty} \lambda_k e^{-\lambda_k t} \phi_k(x) \phi_k(y)$$

is the Mexican hat wavelet defined in Chapter 5. According to [Sun et al. 09], the heat kernel signature is informative, and so is the wavelet signature. This implies that the wavelet signature can faithfully represent the manifold geometry.

Since the heat kernel signature is a monotonically decreasing function of t, it is easy to see that the wavelet signature is always non-negative. Moreover, the all-frequency behavior of wavelet signature satisfies

$$\int_0^\infty \mathrm{WS}(x,t)dt = 1 - \frac{1}{\mu(M)}. \tag{9.20}$$

Proof: The proof is straightforward:

$$
\begin{aligned}
\int_0^\infty \mathrm{WS}(x,t)dt &= \int_0^\infty \sum_{k=0}^\infty \lambda_k e^{-\lambda_k t}\phi_k^2(x)dt \\
&= \sum_{k=0}^\infty \int_0^\infty \lambda_k e^{-\lambda_k t}dt\,\phi_k^2(x) \\
&= \sum_{k=0}^\infty e^{-\lambda_k t}\phi_k^2(x)\Big|_\infty^0 \\
&= \sum_{k=0}^\infty \phi_k^2(x) - \lim_{t\to\infty} h_t(x,x) \\
&= 1 - \frac{1}{\mu(M)}
\end{aligned}
$$

by knowing that

$$\lim_{t\to\infty} h_t(x,y) = \frac{1}{\mu(M)}. \qquad \square$$

It implies that the behavior of wavelet signature at some band can be determined by the wavelet signature at the rest other bands. This is an important property, since usually it is time-consuming to obtain short-time wavelet signature or heat kernel signature through eigenfunction expansion, which requires the full eigen-system. By the wavelet signature, we can analyze high-frequency geometry by studying low-frequency wavelet signatures.

The wavelet signature is also informative, similar to the heat kernel signature.

Theorem 9.2 (Wavelet Signature Informative Theorem) *If the eigenvalues of the Laplace-Beltrami operators of two compact manifolds M and N are not repeated, and T is a homeomorphism from M to N, then T is isometric if and only if $\psi_t^M(x,x) = \psi_t^N(T(x),T(x))$ for any $x \in M$ and any $t > 0$.*

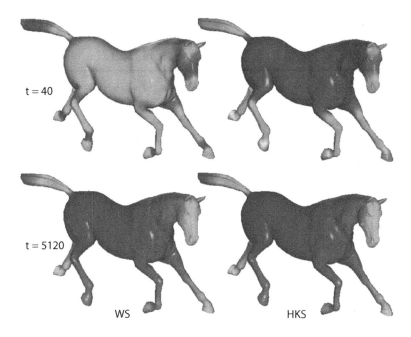

Figure 9.6. Comparison of wavelet signature (WS) and heat kernel signature (HKS) as shape representation. While both of the two are similar at large time, wavelet signature is more distinguishable at small time. See color insert.

Moreover, the trace of the wavelet signature over the domain M is given by

$$\mathrm{WT}(t) = \int_M \mathrm{WS}(x,t)d\mu(x) = \sum_{k=0}^{\infty} \lambda_k e^{-\lambda_k t}, \tag{9.21}$$

which is a direct consequence from the heat trace in Eq.(9.3). Another related representation is the wave kernel signature, introduced in [Aubry et al. 11]. It is a global representation interpreted by energy of particles in quantum mechanics.

In Fig. 9.6, we show a comparison of wavelet signature and heat kernel signature as shape representation. For large scales (large t), the two have very similar value distributions. For small scales (small t), the wavelet signature has more geometric information than the heat kernel signature, which means the wavelet signature represents the shape better.

Interestingly, a recent work in [Kim et al. 13] was inspired by the same idea. It considers the self-effect of the wavelet kernel localized on itself

$\psi_t(x, x)$, and defines a wavelet kernel descriptor with normalization at each frequency t,

$$\text{WKD}(x, t) = \frac{\psi_t(x, x) - \min_y \psi_t(y, y)}{\max_y \psi_t(y, y) - \min_y \psi_t(y, y)}. \tag{9.22}$$

In that work, the signature was used for mesh segmentation and alignment.

9.4.2 Segmentation

The wavelet signature captures shape geometry in a multiscale way; it can be used for most places of the heat kernel signature. Here, we consider one application for shape segmentation. This application is proceeded by simply clustering wavelet signature values. The algorithm is documented in Algorithm 6. Fig. 9.7 (top) shows clustering results on hand models with different topologies. From this example, we see the wavelet signature is sensitive to topology changes that severely modify connecting paths. On the other hand, the segmentation at places other than connectivity changes is still stable. In Fig. 9.7 (bottom), we show multiscale clustering results along different values of t.

Algorithm 6: Shape segmentation by clustering wavelet signatures.

Input: wavelet signature field $\text{WS}(x, t)$, number of clusters N
Output: clusters $\{C_i\}$

Initialize cluster centers $\{c_i\}$;
while *Not convergent* **do**
    for *each* $x \in M$ **do**
        Initialize d_{min} and label l;
        for $i = 0 : N - 1$ **do**
            if $|\text{WS}(x, t) - \text{WS}(c_i, t)| < d_{min}$ **then**
                $d_{min} = |\text{WS}(x, t) - \text{WS}(c_i, t)|$;
                $l = i$;
            end
        end
        Insert x to C_l;
    end
end

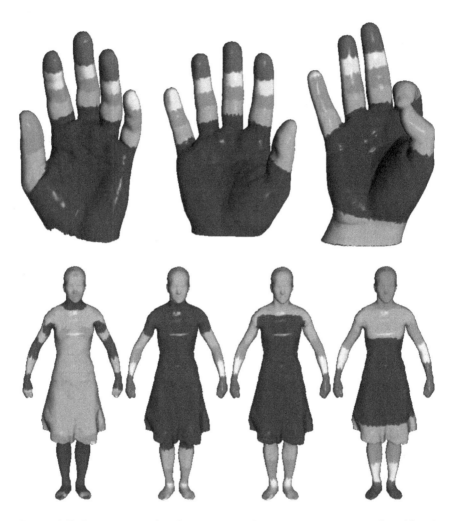

Figure 9.7. Segmentation by clustering wavelet signatures. Top: results of hand models with different topologies: open surface, closed surface, and genus-1 surface. Bottom: multiscale segmentation at different times. See color insert.

10

Geometry Processing

In a broad definition, geometry processing refers to a wide range of operations on 3D meshes: denoising, smoothing, simplification, hole filing, acquisition, and segmentation, to name a few. In this book, we are interested in spectral analysis of 3D shapes. We adopt the concept of "filtering" in [Vallet and Lévy 08], which refers to mesh processing in frequency domain. The filtering of mesh geometry can also be achieved by solving a Poisson equation [Chuang and Kazhdan 11].

From the well known Fourier transform and wavelet transform in signal processing, we can get a picture of how spectral analysis is applied to geometry processing. Generally, filtering by Fourier transform is conducted in frequency domain only, while filtering by wavelet transforms is conducted in both frequency and spatial domain, which can have local operations of geometry processing.

10.1 Fourier Transform

10.1.1 Manifold Harmonics

Processing in spectral domain offers flexible operations for shape geometry. Local areas of curved surfaces are homogeneous to 2D planar patches, where the Euclidean Fourier transform can be applied for spectral processing. In [Pauly and Gross 01], the 2D discrete Fourier transform was applied to local patches constructed from point-sampled geometry. In terms of adapting the Fourier transform on manifolds, basis functions are critical for orthogonally decomposing the space to a series of shape spectra. In [Ben-Chen and Gotsman 05, Karni and Gotsman 00], eigenfunctions of the symmetric Laplacian of the connectivity graph are adopted as a Fourier basis. The processing is achieved by projecting the shape geometry onto an orthonormal basis, which, however, is derived from the mesh topology but not the geometry.

Analogous to Fourier basis for Euclidean domain signals, manifolds have similar orthogonal basis formed by Laplace-Beltrami eigenfunctions

[Lévy 06]. Assume the Laplace-Beltrami operator Δ_M has the eigen-system $\{\lambda_k, \phi_k\}_{k=0}^{\infty}$ with

$$\Delta_M \phi_k = -\lambda_k \phi_k, \tag{10.1}$$

where λ_k is the k-th eigenvalue associated with the eigenvector ϕ_k. The spectrum of Δ_M consists of an increasing positive sequence

$$0 \leq \lambda_0 < \lambda_1 < \cdots.$$

The eigenfunctions $\{\phi_k\}_{k=0}^{\infty}$, also called manifold harmonics, form a complete orthonormal basis in Hilbert space $L^2(M)$. They repetitively oscillate on the manifold with similar behaviors like sine and cosine functions.

Vallet and Lévy [Vallet and Lévy 08] defined the manifold harmonic transform (MHT) of a function $f(x)$ by the inner product

$$\widehat{f}(k) = \langle \phi_k(x), f(x) \rangle, \tag{10.2}$$

where k is related to the "frequency". Since the manifold harmonics basis (MHB) is orthogonal, the inverse transform can be computed by

$$f(x) = \sum_{k=0}^{\infty} \langle \phi_k(x), f(x) \rangle \phi_k(x). \tag{10.3}$$

In [Rong et al. 08], Rong et al. employed spectral decomposition to perform mesh editing on the base domain with low frequencies and reconstruct details with high frequencies.

The Laplace-Beltrami eigenfunctions have nice properties in spectral analysis. However, thousands of eigenfunctions are needed for preserving details of large meshes, which typically demand tremendous computation time and memory. Furthermore, similar to the Fourier basis, the eigenfunctions do not have localization in space domain, which implies that all processes are uniformly operated on the entire manifold. These drawbacks severely limit its application power in mesh editing and geometry processing.

10.1.2   Filtering

One immediate operation of frequency analysis is filtering. A signal can be processed with purposes by frequency filters. As introduced in Chapter 5, the eigenvalue λ_k of the Laplace-Beltrami operator and frequency ω_k of discrete Fourier transform have relation

$$\omega_k = \sqrt{\lambda_k}. \tag{10.4}$$

A frequency filter $\Gamma(\omega_k)$ is a function of amplitudes at each frequency ω_k. A filtering process is a product in frequency domain by simply applying a filter to the manifold harmonic transform of a function [Vallet and

Lévy 08]. For spectral geometry processing, the input is a vector field of vertex coordinates $v(x) = [v_x, v_y, v_z]^T$ of a given mesh, where subscripts x, y, z denote three dimensions of the $\mathbb{R}^3$. By the manifold harmonic transform, the frequency coefficients of vertex coordinates are computed as

$$\widehat{v}(k) = \langle \phi_k(x), f(x) \rangle. \tag{10.5}$$

Computations in the three dimensions are independent. By applying a filter $\Gamma(\omega_k)$, one can obtain a new processed/filtered mesh by

$$v'(x) = \sum_{k=0}^{\infty} \Gamma(\omega_k)\widehat{v}(k)\phi_k(x). \tag{10.6}$$

In practice, a mesh with N number of vertices has at most N eigenvalues and eigenfunctions. As discussed in Chapter 8, we always compute first K of eigenvalues and eigenfunctions, where $K < N$ is a small number (e.g. 300). This actually ignores the high frequency components in the eigen-system. The first K eigenfunctions are incomplete in terms of function space basis. A function $f(x) \in L^2(M)$ is decomposed on first K low frequencies with a residual band of high frequencies

$$f(x) = \sum_{k=0}^{K} \widehat{f}(k)\phi_k(x) + f_h(x), \tag{10.7}$$

where $f_h(x) = f(x) - \sum_{k=0}^{K} \widehat{f}(k)\phi_k(x)$ is the residual high components of $f(x)$. Therefore, the spectral geometry processing in Eq.(10.6) is rewritten as

$$v'(x) = \sum_{k=0}^{K} \Gamma(\omega_k)\widehat{v}(k)\phi_k(x) + \Gamma(\omega_h)f_h(x), \tag{10.8}$$

where $\Gamma(\omega_h)$ is the amplitude at a high frequency ω_h applied to all high frequency components $f_h(x)$.

10.1.3 Example

Fig. 10.1 shows an example of geometry processing by Fourier transform. We design suppression and enhancement filters to simply smooth the shape or enhance the details. One may notice that the enhanced details are not fine. This is because the high frequency component $f_h(x)$ takes all the details that are hard to tell apart. If one wants to achieve operations on very fine details, a large eigenfunction cut-off K is needed.

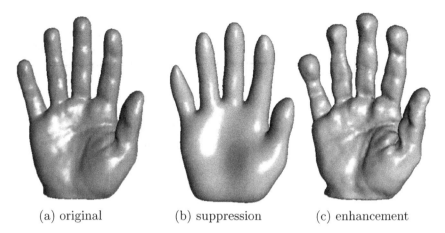

(a) original   (b) suppression   (c) enhancement

Figure 10.1. Geometry processing by the Fourier transform.

10.2   Admissible Diffusion Wavelets

10.2.1   Space-Frequency Filtering

As powerful spectral tools, the admissible diffusion wavelets from Chapter 4 enable geometry analysis and synthesis at different frequencies and places, by filtering their wavelet coefficients. Similarly, vertex coordinates are taken as the input function. Since the vertex coordinate is a 3D vector, its three components are treated independently.

For a given mesh, we compute scaling and wavelet coefficients of its vertex coordinates. Then, we apply filters on the wavelet coefficients to process the geometry. The output mesh can be recovered from a series of wavelet coefficients $\{\mathcal{W}_f(x,j)\}_{j=1}^{J}$ with

$$\mathcal{W}_f(x,j) = \langle \psi_j(x,y), f(y) \rangle \qquad (10.9)$$

and a residual scaling coefficient

$$\mathcal{S}_f(x,J) = \langle \phi_J(x,y), f(y) \rangle \qquad (10.10)$$

using the rapid reconstruction from Chapter 4. It is worthy to notice that the residual term $\mathcal{S}_f(x,J)$ is scaled to the coarsest level, which contains low frequency components. That is, it is easier for the admissible diffusion wavelets to access high frequencies in filtering.

The advantage of wavelets is space-frequency analysis. Unlike the manifold harmonic transform that is an analogue of Fourier transform, the admissible diffusion wavelets can afford local operations during geometry analysis and processing. This can be done instantly by applying a space-frequency filter $\Gamma(x, j)$, which specifies amplitudes at each vertex and each frequency. The processed/filtered mesh can be obtained by

$$v'(x) = \Gamma(x, J)\mathcal{S}_f(x, J) + \sum_{j=1}^{J} \Gamma(x, j)\mathcal{W}_f(x, j). \qquad (10.11)$$

10.2.2 Filter Design

Accordingly, we name $\Gamma(x, j)$ as a global filter if it is equally applied to all $x \in M$, and a local filter if it changes place to place. Global filter $\Gamma(x, j)$ has similar effects with the filter $\Gamma(\omega_k)$ in spectral geometry processing by the manifold harmonic transform. In fact, a global filter $\Gamma(x, j)$ is reduced to $\Gamma(j)$, since it is constant over the spatial variable x.

Local filter $\Gamma(x, j)$ varies over the spatial variable x. Given a mesh M, wavelets construct nested subspaces of function space $L^2(M)$. The space at larger scales is smoothed, and therefore, can be downsampled by ignoring redundant vertices. For the purpose of fast computation, the admissible diffusion wavelets do not downsample scaling and wavelet functions. The function space has not been compressed. Hence, it is reasonable to apply an additional spatial filter in local filter design. A instant spatial filter is the scaling function of admissible diffusion wavelets at level j. For a local filter $\gamma(x, j)$, we apply scaling function $\phi_j(x, y)$ to obtain the final local filter applied to a mesh

$$\Gamma(x, j) = \mathcal{S}_\gamma(x, j) = \langle \phi_j(x, y), \gamma(y, j) \rangle \qquad (10.12)$$

10.2.3 Examples

Fig. 10.2 shows geometry processing using different types of global filters: suppression, enhancement, and band. By this method, the shape geometry can be efficiently filtered, resulting in different effects. The suppression filter smoothes the shape by eliminating the details. On the contrary, the enhancement filter magnifies the details. The two band filters combine the suppression and enhancement in different scales. One approximates the shape at some intermediate scales while keeping fine details. The other enhances the shape at some intermediate scales while smoothing fine details. A comparing result by the Fourier transform is given in Fig. 10.1. The two methods achieve similar effects in global geometry processing, while the admissible diffusion wavelets can further perform local processing.

The local filters are associated with some selected regions, with results shown in Fig. 10.3 and 10.4. In the first example, we carve a letter "S"

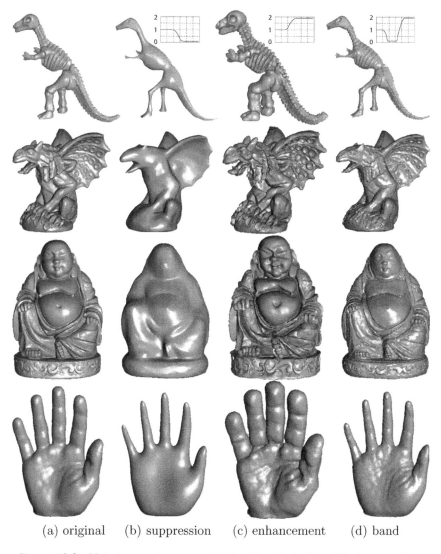

(a) original   (b) suppression   (c) enhancement   (d) band

Figure 10.2. Global geometry processing by the admissible diffusion wavelets.

on a sphere by simply filtering the region selected by hand drawing. It results in a concave shape and a convex shape by the suppression filter and enhancement filter, respectively. The other examples follow the same way by applying space-frequency filters on selected regions, which generates a family of different effects on one shape through a sequence of filtering operations. On the model Gargoyle, selected areas at the wing and the

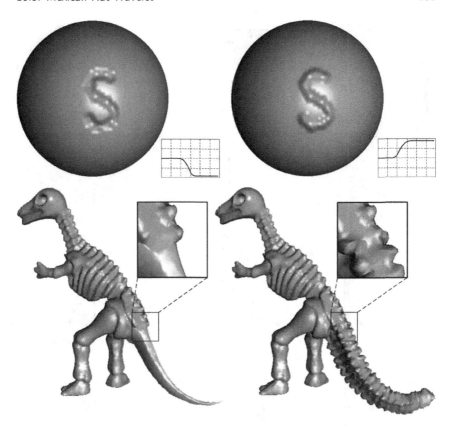

Figure 10.3. Local geometry processing by admissible diffusion wavelets: suppression (left) and enhancement (right) on selected regions.

neck are smoothed, while one horn and the tongue tip are enhanced. The entire processing pipeline can be carried out more efficiently if only the first few frequencies are involved, leading to operations on detail levels.

10.3 Mexican Hat Wavelet

10.3.1 Geometry Processing

The geometry processing can also be achieved by the Mexican hat wavelet. Considering a sequence of discrete wavelets $[\psi_{t_1}, \psi_{t_2}, ..., \psi_{t_J}]$ defined in discrete Mexican hat wavelet (Chapter 5), the wavelet transform of vertex coordinates is given by

$$\mathcal{W}_v(x, t_j) = \langle \psi_{t_j}(x, y), v(y) \rangle. \tag{10.13}$$

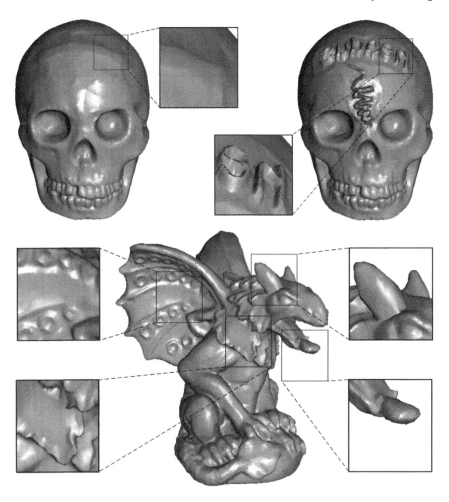

Figure 10.4. More results of local geometry processing by admissible diffusion wavelets: suppression (left) and enhancement (right) on selected regions.

Again, vertex coordinates $v(x)$ is a vector field of three dimensions, which are computed independently.

The output mesh is obtained by the inverse transform of filtered wavelet coefficients and a residual shape,

$$v'(x) = \Gamma(x, t_J)\mathcal{R}_v(x, t_J) + \sum_{j=1}^{J} \Gamma(x, t_j)\mathcal{W}_v(x, t_j). \tag{10.14}$$

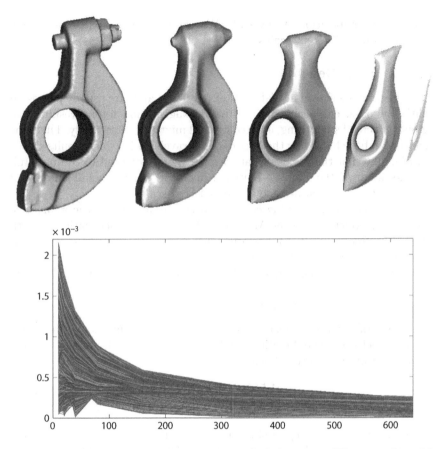

Figure 10.5. Multiscale approximation: residual shapes at different scales with $t_J = 0, 10, 40, 160, 640$ (top) and magnitudes of wavelet transforms at different frequencies (bottom).

Similarly, $\Gamma(x, t_j)$ is space-frequency filter that specifies amplitudes at each vertex and each discrete frequency. The residual component

$$\mathcal{R}_v(x, t_J) = \langle h_{t_J}(x, y), v(y) \rangle \tag{10.15}$$

is the smoothed version at the coarsest level. It contains all low-frequency components of the input.

The residual shape $\mathcal{R}_v(x, t_j)$ is an approximation to $v(x)$ at scale t_j. Therefore, a series of residual shapes compose multiscale approximations of the input. Fig. 10.5 shows residual shapes and magnitudes of wavelet transforms. Details are gradually "peeled off" from the residual shape, and ready for analysis. With the spectral expression, we can rapidly compute spectral decomposition at any scale. Wavelet coefficients extract geometry information at different scales.

10.3.2 Filter Design

Consider results shown in [Vallet and Lévy 08]; shapes are usually processed with low-pass, enhancement, and band exaggeration filters, resulting in global effects of smoothing, enhancing, and mixture, respectively. The same effects can be achieved by the Mexican hat wavelet, with some results shown in Fig. 10.6. For the Mexican hat wavelet, we do not need to compute full eigen-system of the Laplace-Beltrami operator, since a small number of eigenfunctions is accurate enough for selected scales.

Moreover, the processing can be spatially different like the admissible diffusion wavelets, since the Mexican hat wavelet has localization in both space and frequency. It can perform space-frequency analysis on geometry. This allows local operations of spectral processing on geometry. Specifically, we design space-frequency filters by considering additional spatial smoothing

$$\Gamma(x,j) = \langle h_{t_j}(x,y), \gamma(y,j) \rangle. \tag{10.16}$$

Example of space-frequency filters along with some lower-frequency Laplace-Beltrami eigenfunctions and some results are shown in Fig. 10.7 and Fig. 10.8. Precisely, a linear filter along $\phi_k(x)$

$$\gamma(x,t_j) = \frac{1.5(\phi_k(x) - \min(\phi_k(x)))}{\max(\phi_k(x)) - \min(\phi_k(x))} + 0.5, \tag{10.17}$$

is applied to all $\mathcal{W}_v(x,t_j)$. Using these filters, the input shape is gradually morphed from smoothing to enhancement. Other types of filters can also be designed according to specific applications.

Figure 10.6. Global geometry processing by the Mexican hat wavelet, from left to right: original shapes, enhanced shapes, and smoothed shapes.

Figure 10.7. Local geometry processing by the Mexican hat wavelet, from left to right: original shapes, and two columns of filtered shapes. The filters are linear to some lower-frequency Laplace-Beltrami eigenfunctions.

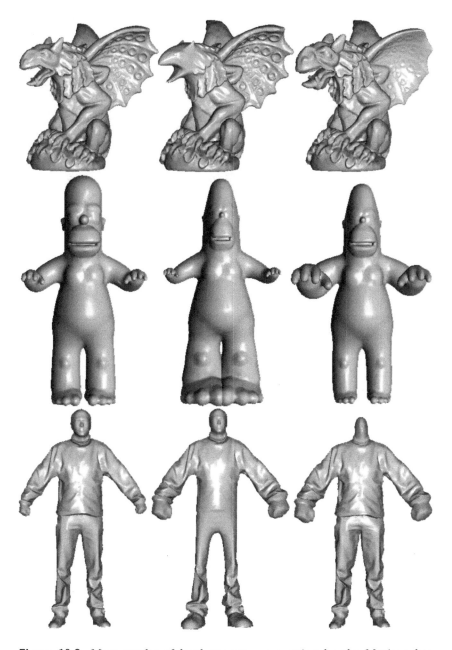

Figure 10.8. More results of local geometry processing by the Mexican hat wavelet, from left to right: original shapes, and two columns of filtered shapes. The filters are linear to some lower-frequency Laplace-Beltrami eigenfunctions.

11

Feature Definition and Detection

Shape feature plays a critical role in many graphics and visualization tasks. There are many strategies to define features. Among them, wavelet transform is widely used, because it extracts detailed information of a function at different scales. In image processing, the well known scale-invariant feature transform (SIFT) [Lowe 04] is a wavelet-based method. SIFT boosted many high-level problems in computer vision with robust features. One reason behind the scene is that human vision is very sensitive to second-order derivatives. The difference of Gaussians, as a wavelet, is believed to mimic how neural processing in the retina of the eye extracts details from images destined for transmission to the brain [Young 87].

This chapter focuses on wavelet transform as an effective method for feature definition and detection. Features found as zero-crossings in second-order derivative of a scale space are multiscale, resilient, and representative. It first introduces saliency visualization as a map contains feature information. Then it revisits the classical definition of features as local extrema of differences in a scale space. At last, it demonstrates some feature detection results by wavelet tools introduced in this book.

11.1   Saliency Visualization

Saliency visualization is a study to visualize saliency information of 3D shapes. It can be taken as the first step of feature detection, by showing intermediate results for applications needing the information. In [Lee et al. 05], Lee et al. defined the saliency map as the difference of two Gaussian convoluted scalar fields of a curvature map, and the mesh saliency as the nonlinear combination of the saliency maps in multiple scales. As a useful tool, it has been applied to mesh simplification [Lee et al. 05], volume visualization [Kim and Varshney 06], feature matching [Castellani et al. 08], visual perception [Kim et al. 10], etc.

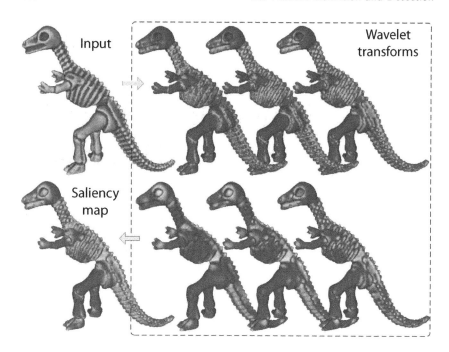

Figure 11.1. A saliency map is the nonlinear sum of its wavelet coefficients. See color insert.

Wavelets, for example the admissible diffusion wavelets, have similar effects for shape representation. For a given function $f(x)$, its wavelet coefficients encode the saliency information of multiscale details. Analogous to [Lee et al. 05], the saliency map is computed by the nonlinear sum of wavelet coefficients at different scales,

$$\Lambda(f) = \sum_j \alpha_j |\mathcal{W}_f(j)|, \qquad (11.1)$$

where

$$\alpha_j = (\max(|\mathcal{W}_f(j)|) - \text{mean}(|\mathcal{W}_f(j)|))^2 \qquad (11.2)$$

is a coefficient. Fig. 11.1 shows the workflow of computing the saliency map by using the admissible diffusion wavelets. To avoid being affected by noise perturbation, the first two levels of $\mathcal{W}_j$ are ignored. The wavelet coefficients capture the changes of the function at different scales. Low-frequency wavelets capture small-scale changes, while high-frequency ones capture large-scale changes.

Fig. 11.2 shows the saliency maps of mean curvature maps on two deformed shapes of the Armadillo, where the top row is the result of [Lee et al. 05] and the bottom row is done by the admissible diffusion wavelets.

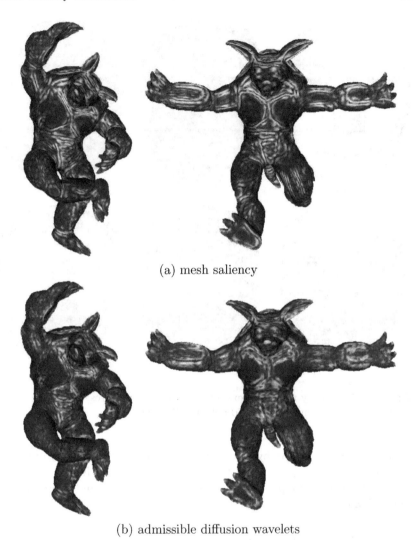

(a) mesh saliency

(b) admissible diffusion wavelets

Figure 11.2. Saliency visualization by (a) mesh saliency [Lee et al. 05] and (b) admissible diffusion wavelets.

The original mesh saliency in [Lee et al. 05] is computed by an Euclidean Gaussian. In general, the two methods both adopt the *difference-of-Gaussian* operator, and therefore, they have similar effects in saliency visualization. The setting of the mesh saliency in multiscale Gaussian is more empirical, while the admissible diffusion wavelets have a rigorous construction by means of wavelets. They have some differences in counting saliency components at different scales. In [Hou and Qin 13], we also

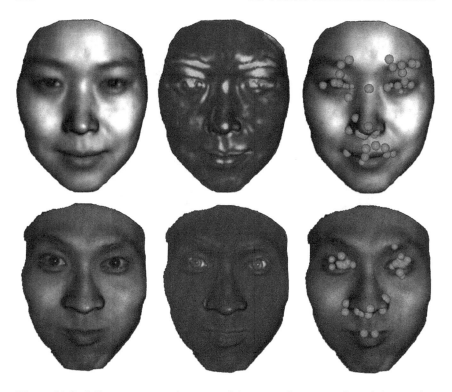

Figure 11.3. Saliency maps and extracted features of textures, from left to right: input textures, saliency maps, and detected features.

compute saliency maps by the admissible diffusion wavelets from other functions on the shape, such as texture, and vertex coordinates. Fig. 11.3-11.5 show saliency maps of texture on meshed surfaces and coordinates of point clouds.

11.2   Feature Definition

11.2.1   Related Work

Feature definition determines the qualities and properties of found features, and therefore is crucial to applications of shape feature. A prevailing method in image processing is to define features through a scale space based on diffusion [Lowe 04]. The scale space is obtained by continually smoothing a function by some diffusion kernel.

Many attempts have been made to prompt it on manifold surfaces. An intuitive idea is to flatten surfaces to 2D images via parameterization,

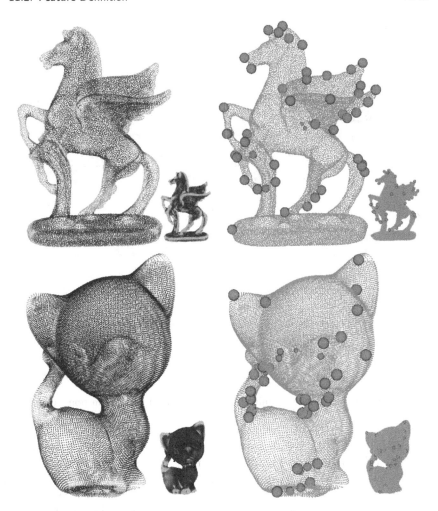

Figure 11.4. Saliency visualization and feature detection on point clouds. See color insert.

and then compute scale-space features of geometric attributes such as normal [Novatnack and Nishino 07] and curvature [Hua et al. 08]. The parameterization itself, however, suffers from unpredictable changes of topology and boundary, accompanied by domain cutting and shape distortion. In [Wu et al. 08], texture was projected to the tangent plane to locally flatten the surface. This method, however, was designed for surfaces with simple geometric shape such as walls.

Purely derived from geometry, some other work constructed scale space directly on the 3D surfaces evolving the scale domain information.

Figure 11.5. Saliency visualization and feature detection on point clouds.

In [Pauly et al. 03], a scale space was formulated via surface variation on point-sampled surfaces. Line-type features were extracted by a multiscale classification operator that approximates the surface at different scales. In [Zou et al. 09], an intrinsic geometric scale space of 3D surfaces was proposed for extracting scale-dependent saliency. Using Ricci Flow, the surface gradually changes its curvature via shape diffusion. This scale space, therefore, is invariant to conformal deformation. For scale space on surfaces, Lee et al. [Lee et al. 05] adopted 3D Gaussian convolution of curvature maps to compute mesh saliency. The 3D Gaussian scale space is easy to compute; nevertheless, it is based on Euclidean distance and only feasible for rigid objects. To improve this, a geodesic scale space [Zou et al. 08] was introduced using geodesic-based Gaussian convolution. The cost of computing geodesics, however, is extremely high as the scale increases. Recently, the concept of scalar fields defined on 3D surface has been proposed [Zaharescu et al. 09], which nicely combines the photometric and geometric characteristics together. However, it is not accurate for large scales, where the fundamental solution to the heat equation is not a Gaussian.

Therefore, challenges are still found upon theoretical accuracy, computational efficiency, parameter selection, etc.

Shape features can also be described by global properties, such as Laplace-Beltrami eigenvalues [Reuter et al. 06] and eigenfunctions [Rustamov 07]. Nevertheless, global properties are easily affected by changes of topology and geometric boundary. Recently, Sun et al. [Sun et al. 09] proposed the heat kernel signature to describe and detect manifold features. It starts to gain more popularity in the state-of-the-art for its merits in stableness, multi-scale, isometric invariance, and informativeness. It has been successfully applied to shape matching and registration [Dey et al. 10, Ovsjanikov et al. 10]. In [Patanè and Falcidieno 10], convolution based on heat kernels was used to find multi-scale features. Besides, some high-level features, i.e. curves and patches, have emerged in recent work [Kim et al. 09, Sunkel et al. 11, Wang et al. 11b].

11.2.2 Wavelet Feature

Features are oftentimes interpreted as local extrema in first-order derivative of a scale space, i.e. zero-crossings in the second-order derivative. The essence of multiscale feature is to construct multiscale approximations for a function under diffusion, and find maximum changes in the scale space. This scheme is referred to as the scale-invariant feature transform (SIFT) [Lowe 04]. For a given image $I(x, y)$, a diffused version at scale σ is given by

$$L(x, y, \sigma) = G(x, y, \sigma) * I(x, y), \tag{11.3}$$

where $*$ denotes the convolution operator, and $G(x, y, \sigma)$ is a Gaussian kernel at scale σ

$$G(x, y, \sigma) = \frac{1}{2\pi\sigma^2} e^{\frac{x^2+y^2}{2\sigma^2}}. \tag{11.4}$$

A series of diffused versions compose a scale space of $I(x, y)$. The difference of Gaussian (DoG) image $D(x, y, \sigma)$ is computed from the difference of two nearby scales separated by a constant multiplicative factor k:

$$D(x, y, \sigma) = L(x, y, k\sigma) - L(x, y, \sigma). \tag{11.5}$$

The difference-of-Gaussian function provides a close approximation to the scale-normalized Laplacian of Gaussian $\sigma^2 \Delta G$. As we know, the Gaussian kernel is the heat kernel in Euclidean space. According to the heat equation parameterized in terms of σ rather than $t = \sigma^2$

$$\frac{\partial G(x, y, \sigma)}{\partial \sigma} = \sigma \Delta G(x, y, \sigma). \tag{11.6}$$

Considering wavelets can be defined as the negative first-order derivative of heat kernel, it indicates the difference-of-Gaussian transform is normalized wavelet transform. SIFT features are actually local extrema of wavelet coefficients.

This schema can be applied to functions on manifolds by the aforementioned wavelets. Generally, given a function $f(x) \in L^2(M)$ and wavelets $\psi_t(x, y)$, the wavelet transform is computed by

$$\mathcal{W}_f(x, t) = \langle \psi_t(x, y), f(y) \rangle, \tag{11.7}$$

where the convolution operation is replaced by inner product on manifolds. The function $f(x)$ could be coordinates, texture, curvature map, density, etc. The wavelet transforms capture details of an input function at different scales. Features are defined as local extrema of wavelet coefficients.

11.3   Feature Detection

In [Lowe 04], SIFT features are detected in a 3D space with scale as the third dimension. This is to reduce the number of detected features by only leaving those more remarkable. We adopt this scheme for detection of wavelet features. Specifically, a point x is recognized as a feature if it is an extremum in its local neighborhood in a complex space-frequency domain. In spatial domain, the neighborhood of a vertex contains, for example, its 2-ring neighbors for meshed surface or 10 nearest neighbors for point clouds. In frequency domain, the neighborhood is two adjacent levels of scales (frequencies).

The algorithm is documented in Algorithm 7. It takes an input function $f(x)$ and a series of wavelets $\{\psi_t(x, y)\}$ at different scales. It filters out points with small magnitudes of wavelet coefficients. As for detection, it checks a candidate by comparing with its neighbors in both spatial domain and frequency domain. Candidates that are local maxima or minima are selected as features.

Below we show feature detection results by wavelet tools introduced in this book.

11.3.1   Diffusion-Wavelets Feature

In Fig. 11.6, extracted features of vertex coordinates are shown as green balls with their sizes corresponding to their correct scales. The threshold θ controls the number of features, according to their values of saliency. Features with greater values of wavelet coefficients are more salient, thus,

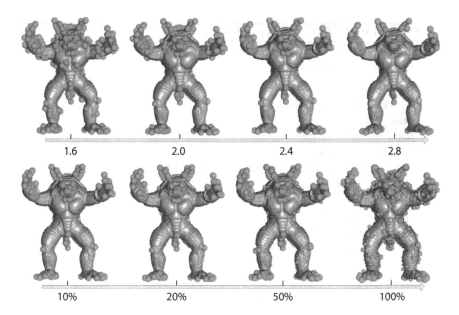

Figure 11.6. Feature detection on the Armadillo with different thresholds and noise levels. Features are rendered by balls with their sizes representing correct scales. Large-scale features are more stable than small-scale features.

more stable. Fig. 11.6 also shows an experiment with noisy input. We add Gaussian noise with increasing standard deviation σ of mean edge length in normal directions. Large-scale features appear to be more stable than small-scale features. Even with large noise (50%), one can still find stable features at the larger scales, which are exactly the same as that on the noise-free shape.

Fig. 11.3 shows a result with texture input. The admissible diffusion wavelets can rapidly find rich features from the scale space. Textures on meshed surfaces have information other than shape geometry, which is capable of finding features in areas with indistinctive shape but rather rapidly-changing texture. The wavelet transform of admissible diffusion wavelets extracts multiscale details of the input function, and finds features as zero-crossings of second-order derivatives in its multiscale representation.

For point clouds in Fig. 11.4 and Fig. 11.5, we directly use point coordinates as input functions $v(x)$. The wavelet coefficients in three dimensions are consolidated together by norm $|\mathcal{W}_v(j,x)|_2$. In these experiments, we choose not to apply feature filters, for instance edge filter and boundary filter that could eliminate features on edges and boundaries, respectively.

Algorithm 7: Wavelet feature detection.

Input: function $f(x)$, wavelets $\psi_t(x,y)$, scales $\{t_0, t_1, \cdots, t_J\}$
Output: feature sets F

for $j = 0 : J$ **do**
  | $\mathcal{W}_f(x, t_j) = \langle \psi_{t_j}(x,y), f(y) \rangle$;
end
for $j = 1 : J - 1$ **do**
  | **for** $x \in M$ **do**
      | **if** $|\mathcal{W}_f(x, t_j)| < \text{mean}(|\mathcal{W}_f(x, t_j)|)$ **then**
        | continue;
      end
      $state = 0$;
      for $y \in N(x)$ **do**
        $state\_tmp = 0$;
        if $\mathcal{W}_f(x, t_j) > \mathcal{W}_f(y, t_j)$ **then**
          | $state\_tmp = 1$;
        else if $\mathcal{W}_f(x, t_j) < \mathcal{W}_f(y, t_j)$ **then**
          | $state\_tmp = -1$;
        end
        if $state == 0$ **then**
          | $state = state\_temp$;
        else if $state * state\_tmp <= 0$ **then**
          | break;
        end
      end
      for $d \in \{-1, 1\}$ **do**
        $state\_tmp = 0$;
        if $\mathcal{W}_f(x, t_j) > \mathcal{W}_f(x, t_{j+d})$ **then**
          | $state\_tmp = 1$;
        else if $\mathcal{W}_f(x, t_j) < \mathcal{W}_f(x, t_{j+d})$ **then**
          | $state\_tmp = -1$;
        end
        if $state == 0$ **then**
          | $state = state\_temp$;
        else if $state * state\_tmp <= 0$ **then**
          | break;
         end
      end
      Insert x into F;
  end
end

Figure 11.7. Mexican hat wavelet transform of vertex coordinates at different scales. See color insert.

Depending on the purpose of feature extraction in different practice, these filters can be easily introduced and take appropriate actions.

11.3.2 Mexican-Hat-Wavelet Feature

The Mexican hat wavelet can also be used for feature detection by Algorithm 7. Particularly, when inputs are vertex coordinates $v = [v_x, v_y, v_z]$ of a meshed surface, the half normal of wavelet coefficients in three dimensions

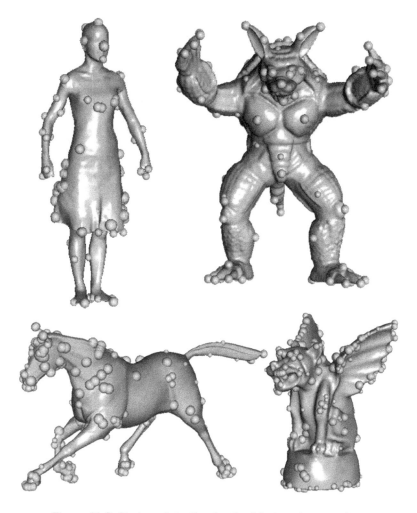

Figure 11.8. Feature detection by the Mexican hat wavelet.

of $\mathbb{R}^3$ is convergent to the mean curvature as $t \to 0$.

$$\lim_{t \to 0} \frac{1}{2} |\mathcal{W}_v(x, t)| = \frac{1}{2} |\Delta_M \cdot v| = \kappa_H \qquad (11.8)$$

This is because the short-time Mexican hat wavelet converges to the Laplace-Beltrami operator, and half normal of $\Delta_M \cdot v$ is the mean curvature.

Fig. 11.7 (top) shows continuous wavelet transform maps of vertex coordinates at $t = 10$ and $t = 400$ on the Armadillo. Fig. 11.7 (bottom) shows

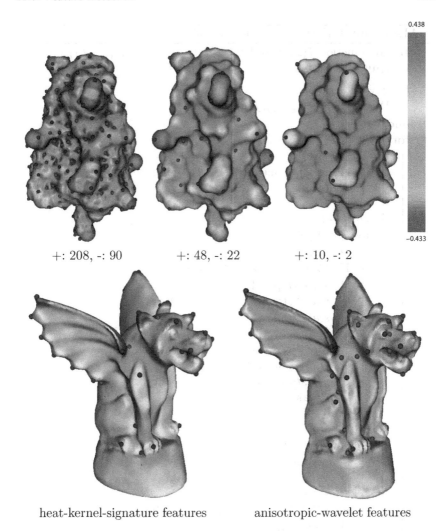

0.438

+: 208, -: 90   +: 48, -: 22   +: 10, -: 2

−0.433

heat-kernel-signature features   anisotropic-wavelet features

Figure 11.9. Top: Anisotropic features detected at different scales on the Blob. The convex (+) and concave (-) features are highlighted in blue and red balls, respectively. Bottom: Comparison of heat-kernel-signature features (left) and anisotropic-wavelet features (right) detected on the Gargoyle. See color insert.

some more wavelet transforms of the shape geometry at different scales. Fig. 11.8 gives some results of feature detection, where features are shown as balls whose sizes depend on their scales.

Algorithm 8: Feature detection by anisotropic wavelets.

Input: function $f(x)$, wavelets $\psi_t^e(x, y)$, scales $\{t_0, t_1, ..., t_J\}$
Output: convex feature sets F_v, concave feature sets F_c

for $j = 0 : J$ **do**
  | $\mathcal{W}_f^e(x, t_j) = \langle \psi_{t_j}^e(x, y), f(y) \rangle$;
end
for $j = 1 : J - 1$ **do**
  | **for** $x \in M$ **do**
  | | **if** $|\mathcal{W}_f^e(x, t_j)| < \text{mean}(|\mathcal{W}_f^e(x, t_j)|)$ **then**
  | | | continue;
  | | **end**
  | | $state = 0$;
  | | **for** $y \in N(x)$ **do**
  | | | $state\_tmp = 0$;
  | | | **if** $\mathcal{W}_f^e(x, t_j) > \mathcal{W}_f^e(y, t_j)$ **then**
  | | | | $state\_tmp = 1$;
  | | | **else if** $\mathcal{W}_f^e(x, t_j) < \mathcal{W}_f^e(y, t_j)$ **then**
  | | | | $state\_tmp = -1$;
  | | | **end**
  | | | **if** $state == 0$ **then**
  | | | | $state = state\_temp$;
  | | | **else if** $state * state\_tmp <= 0$ **then**
  | | | | break;
  | | | **end**
  | | **end**
  | | **for** $d \in \{-1, 1\}$ **do**
  | | | $state\_tmp = 0$;
  | | | **if** $\mathcal{W}_f^e(x, t_j) > \mathcal{W}_f^e(x, t_{j+d})$ **then**
  | | | | $state\_tmp = 1$;
  | | | **else if** $\mathcal{W}_f^e(x, t_j) < \mathcal{W}_f^e(x, t_{j+d})$ **then**
  | | | | $state\_tmp = -1$;
  | | | **end**
  | | | **if** $state == 0$ **then**
  | | | | $state = state\_temp$;
  | | | **else if** $state * state\_tmp <= 0$ **then**
  | | | | break;
  | | | **end**
  | | **end**
  | | **if** $state > 0$ **then**
  | | | Insert x into F_v;
  | | **else**
  | | | Insert x into F_c;
  | | **end**
  | **end**
end

11.3.3   Anisotropic-Diffusion-Wavelet Feature

Given a heat field constructed using local geometry quantities, such as normal signature in Chapter 6, we can detect multi-scale features by analyzing the anisotropic diffusion wavelets. Recall that the anisotropic diffusion wavelet $\psi_t^e(x, y)$ in Chapter 6 is based on an edge-weighted heat kernel, which is therefore anisotropic. Considering the normal-controlled coordinates are distinguishable at convex and concave vertices, one can even separate convex and concave features by slightly modifying Algorithm 7 to Algorithm 8.

For small scales, the detected features depend on more local geometry information. For large scales, more global information is taken into account. Fig. 11.9 (top) shows the features detected at different scales. As scale increases, the heat field becomes smooth gradually. The features detected in different scales can well depict the shape information in a multi-scale sense. Fig. 11.9 (bottom) shows a comparison between anisotropic-wavelet features and heat-kernel-signature [Sun et al. 09] features. The heat kernel cannot distinguish convex and concave features, and has a limitation on small-scale features at short time even though all the eigenfunctions are used.

12

Shape Matching, Registration, and Retrieval

By adopting the categorization in computer vision, shape representation, geometry processing, and feature detection can viewed as low-level problems, as they are fundamental to other applications. This chapter presents some extended middle-level to high-level problems facilitated by the low-level tools. It further demonstrates how wavelet tools can be involved in solving many problems in computer graphics and computer vision.

12.1   Shape Matching

Shape matching is to match selected points on shapes, with purposes in shape recognition and retrieval. In graphics, the topic of shape matching started to gain momentum recently, which is often performed by building a map between two sets of points, or searching common subgraphs of two graphs.

Lipman and Funkhouser [Lipman and Funkhouser 09] utilized conformal mapping and a voting strategy to match samples. In [Dubrovina and Kimmel 10], the eigenfunctions of the Laplace-Beltrami operator were employed for shape matching, which are global functions defined on the entire shape. In [Ovsjanikov et al. 10], Ovsjanikov et al. defined the heat kernel map as heat kernels from a fixed reference point, and applied it to shape matching using a simple nearest neighbor search. They also extended the one-point method to multiple reference points. In [Dey et al. 10], a shape was mapped to a feature space, and represented by feature vectors. The matching of two shapes turns into scoring the shape pair using feature vectors. In [Sun et al. 10], a correspondence map was measured by the intersection configuration distance. Bronstein et al. adopted diffusion distance as an isometry-invariant metric in shape recognition [Bronstein and Bronstein 11] and shape matching [Bronstein et al. 10b]. In [Raviv et al. 11], an affine invariant diffusion geometry was proposed for matching deformable shapes, which is invariant to squeeze and shear transformations.

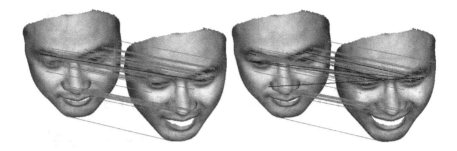

Figure 12.1. Nearest neighbor matching results from [Hou and Qin 10].

In computer vision, this problem can also be well modeled by the graph matching with pair-wise or high-order geometric compatibilities. Leordeanu and Hebert [Leordeanu and Hebert 05] proposed the spectral matching method that constructs a graph by taking each candidate correspondence as a point. Correct correspondences were found by the eigenvector with the largest eigenvalue of the affinity matrix, whose entries are pair-wise geometric relations. In [Cour et al. 06], Cour et al. improved the second-order graph matching by a spectral relaxation scheme and a normalization procedure. Torresani et al. [Torresani et al. 08] proposed an optimization technique of high-order graph matching using dual decomposition. Recently, Duchenne et al. [Duchenne et al. 09] extended the spectral matching to the affinity tensor with third-order potentials, which improves the robustness but requires heavy computational load. In [Zeng et al. 10], a 3D shape was flattened to a 2D image using conformal map in order to adopt the high-order graph matching. However, the surface flattening has strict requirements on topology.

Here we consider the shape matching problem as matching two feature sets. Given two sets of feature points F_1 and F_2, the problem is to find reliable matches $\{i = (i_1, i_2) \mid i_1 \in F_1, i_2 \in F_2\}$. Features are detected on deformable shapes, which bring many challenges to this problem.

12.1.1 Nearest Neighbor

A simple method of matching two feature sets is nearest neighbor. For a point $i_1 \in F_1$, it finds the most similar point $i_2 \in F_2$, as shown in Algorithm 9, where $s(i_1, i_2)$ is a similarity function of two points i_1 and i_2, measured by some distance in feature space under some feature representation [Lipman and Funkhouser 09, Ovsjanikov et al. 10, Hou and Qin 10]. The found matching should be discriminative, i.e. the similarity of the closest neighbor is θ or more to that of the second-closest neighbor.

Fig. 12.1 shows some matching results by nearest neighbor adopted from [Hou and Qin 10], where light lines indicate correct matches and dark

---

Algorithm 9: Feature matching by nearest neighbor

Input: two sets of points F_1 and F_2, similarity function $s(i_1, j_2)$,
      parameter θ

Output: matched points $\{i = (i_1, i_2) \,|\, i_1 \in F_1, i_2 \in F_2\}$

for $i_1 \in F_1$ **do**

    Initialize s_max_1, s_max_2;

    for $j_2 \in F_2$ **do**

        if $s(i_1, j_2) < s\_max_1$ **then**

            $s\_max_1 = s(i_1, j_2)$;

            $i_2 = j_2$;

        else if $s(i_1, j_2) < s\_max_2$ **then**

            $s\_max_2 = s(i_1, j_2)$;

        end

    end

    if $s\_max_1 > \theta \cdot s\_max_2$ **then**

        Insert (i_1, i_2) to output;

    end

end

---

lines indicate mismatches. When undergoing large deformation, the nearest neighbor produces unreliable results.

There are two major issues with this method. Firstly, it is a brute-force method that checks every pairs of two feature sets. It runs very slow with time complexity $O(V^2)$, with V denotes the number of vertices. An improvement is to use the approximate nearest neighbor. Secondly, it only considers point-wise similarity, but ignores geometric compatibility. The geometric compatibility means that points on one shape should have compatible geometric relations with their corresponding points (if such points indeed exist) on the other shape.

12.1.2 Spectral Matching

The spectral matching was introduced by [Leordeanu and Hebert 05], and was adopted for shape feature matching in [Hou and Qin 12b].

Assume we have two feature sets F_1 and F_2 on two shapes. A pair $i = (i_1, i_2)$ denotes a candidate match with $i_1 \in F_1$ and $i_2 \in F_2$. For a candidate pair $i = (i_1, i_2)$, its binary assignment is given by

$$x(i) = \begin{cases} 1 & \text{if } i \text{ is chosen} \\ 0 & \text{otherwise} \end{cases}, \text{ with } \sum_{i_2} x(i) \leq 1. \qquad (12.1)$$

To save computation and reduce searching space, only pairs with high point-wise similarities will be selected into a candidate set

$$C = \{i \mid i_1 \times i_2 \in F_1 \times F_2, \ s(i_1, i_2) > \epsilon_s\}, \qquad (12.2)$$

where ϵ_s is a threshold for candidate pair selection. The candidate set C will be the search space. Assuming each feature has constant number of matching candidates, the size of search space is linear to the number of features.

The matching problem now turns to solve an objective function

$$J(x) = x^T A x, \qquad (12.3)$$

where A is affinity matrix that measures similarities of all feature pairs. The matching problem is equivalent to finding an assignment x that maximizes the objective function.

Given a distance function $d(i_1, j_1)$ that measures the distance between two points on one shape, with an example in [Hou and Qin 12b], the relative compatibility of two pairs i, j can be modeled by

$$c(i, j) = \frac{|d(i_1, j_1) - d(i_2, j_2)|}{d(i_1, j_1) + d(i_2, j_2)}. \qquad (12.4)$$

The affinity matrix $A(i, j)$, which represents relations of all candidate pairs, is constructed as follows:

1. The entry $A(i, i)$ is the affinity of the correspondence itself, which is computed by

$$A(i, i) = e^{-\frac{s(i_1, i_2)^2}{2\sigma_s^2}}, \qquad (12.5)$$

where $s(i_1, i_2)$ is a point-wise similarity function of two points, and σ_s is a parameter that can be set as $\sigma_s = \epsilon_s$.

2. The affinity $A(i, j)$ represents the geometric compatibility of two candidate correspondences, defined as

$$A(i, j) = e^{-\frac{c(i, j)^2}{2\sigma_c^2}}, \qquad (12.6)$$

where $c(i, j)$ is a function measuring the compatibility of two candidate pairs, and the parameter σ_c can be set as 0.1.

3. If two candidate correspondences i and j are conflicting (e.g., $i_1 = j_1$ and $i_2 \neq j_2$), we let $A(i, j) = 0$.

Usually, the objective function $J(x)$ is relaxed to

$$J'(x) = \frac{x^T A x}{x^T x}. \qquad (12.7)$$

Algorithm 10: Binary projection.

Input: optimal solution x^* in Eq. (12.8)
Output: binary assignment x_b

sort x^*;
while x^* *has element* $> \epsilon_b$ **do**
   find the greatest element p in x^*;
   $x_b(p) = 1$;
   for *element* q *in* x^* **do**
      if q *and* p *are conflict* **then**
         $x^*(q) = 0$;
      end
   end
   $x^*(p) = 0$;
end

And the optimal solution maximizes the objective function:

$$x^* = \arg\max_x \left(\frac{x^T A x}{x^T x} \right). \tag{12.8}$$

It can be solved by computing the leading eigenvector of A. The final assignment is the binary projection of x^*, subject to conflicting constraints. This binary projection algorithm is given in Algorithm 10, where $\epsilon_b = 0.1$ is a threshold to select matches with strong reliabilities.

Some results of spectra matching are shown in Fig. 12.2. The spectral matching finds less matches compared with the nearest neighbor matching. However the found matches are reliable to very large deformation, for example faces from different persons in Fig. 12.2. Features that may not have high point-wise similarities are matched collectively as a graph.

12.1.3 Graph Matching

The spectral matching is a graph matching with second-order terms. High-order graph matching [Duchenne et al. 09, Torresani et al. 08] has been shown as a powerful and stable method to match point sets against outliers. It considers geometric compatibilities of feature tuples, and formulates the objective function using high-order potentials of their affinities. Geometric relations among features are extremely important on deformable shapes, and they are more reliable than single feature points towards shape matching. Therefore, graph matching directly on manifold without surface flattening is highly desirable with pressing needs.

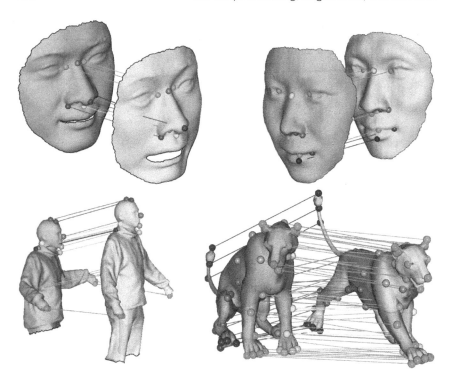

Figure 12.2. Spectral matching results from [Hou and Qin 12b]. See color insert.

A nice formulation of high-order graph matching is the tensor formulation [Duchenne et al. 09]. In [Hou et al. 12b], it was employed for matching o shape features. Consider a tuple of three candidate matches (i, j, k) without conflicts, i.e., $i_1 \neq j_1 \neq k_1$ and $i_2 \neq j_2 \neq k_2$. As shown in Fig. 12.3, they can form two "triangles" by some distance function.

The tuple of candidate matches is then embedded into a 3D space by three angles of this "triangle". The distance in the embedded space is given by

$$d_\theta(i, j, k) = \|\theta_{i_1, j_1, k_1} - \theta_{i_2, j_2, k_2}\|_2, \qquad (12.9)$$

where θ_{i_1, j_1, k_1} is a vector comprising three angles of the triangle formed by points i_1, j_1, k_1, and $\|.\|_2$ denotes the l^2-norm. The affinity of the tuple (i, j, k) without conflicts is defined as

$$h_{i,j,k} = e^{-d_\theta(i,j,k)^2/\sigma}, \qquad (12.10)$$

where σ is a parameter, which can be set as $\sigma = \text{mean}(d_\theta)$. For tuples with conflicts, we let their affinities equal to zero. The high-order score of

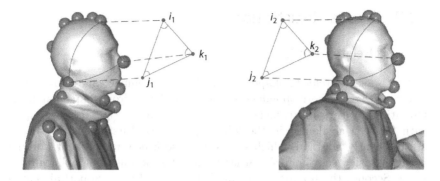

Figure 12.3. Third-order potentials of manifold matching. Three candidate matches (i, j, k) form two "triangles" on two manifolds.

assignment X is defined as

$$\text{score}(X) = \sum_{i,j,k} h_{i,j,k} X_{i_1,i_2} X_{j_1,j_2} X_{k_1,k_2}. \tag{12.11}$$

We can rewrite the score using tensor notation, given by

$$\text{score}(X) = H \otimes_1 X \otimes_2 X \otimes_3 X, \tag{12.12}$$

where $\otimes_d$ denotes the tensor product in d dimension. Here, H is a *3rd*-order tensor with entries $h_{i,j,k}$ defined in Eq. (12.10). The final results are obtained according to their matching scores subject to conflict constraints.

For the high-order optimization in Eq. (12.12), one can use the tensor power iteration with l^1-norms of columns. For more details of this algorithm, we refer readers to [Duchenne et al. 09]. Fig. 12.4 shows some graph matching results adopted from [Hou et al. 12b].

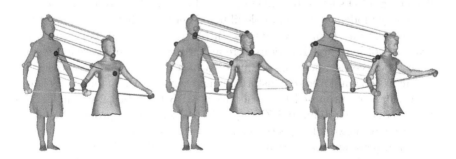

Figure 12.4. Graph matching results from [Hou et al. 12b].

12.2   Shape Registration

12.2.1   Related Work

In this book, we refer to shape registration as a compact matching of two shapes with dense correspondences, which is different from shape matching. Oftentimes, we are addressing dense registration of nonrigid partial shapes, with foreseen challenges in the following aspects. First, dynamic shapes are frequently nonrigid, which forces us to seek new methods of working with intrinsic "coordinates" rather than extrinsic ones in Euclidean domain. Second, the fragments acquired from the object are partial shapes with overlapping regions and changing boundaries. Therefore, the globally-defined coordinates are not applicable, including the shape-DNA [Reuter et al. 06], the global point signature [Rustamov 07], mean value coordinates [Ju et al. 05], and harmonic coordinates [Joshi et al. 07], to name just a few. Third, the output correspondences must be dense, which is compounded by more difficulties, including accuracy of registration, geometric and topological compatibility, computational efficiency, etc.

For articulated objects, piece-wise rigid transformation [Chang and Zwicker 08, Huang et al. 08] was adopted to segment the surface to rigid sub-parts, lacking accuracy for nonrigid deformation. For complete nonrigid shapes, global map has been widely used in previous methods. Kraevoy and Sheffer [Kraevoy and Sheffer 04] used cross-parameterization to establish a bijective mapping between two surfaces through a common base domain. The mapping from a surface to its base domain is initialized by mean value parameterization, and the triangular interiors in the base domain are registered using barycentric coordinates. Manifold harmonics are also used to register shapes with isometric deformation by spectral embeddings [Jain et al. 07, Mateus et al. 08, Reuter et al. 09, Ruggeri et al. 10], which are limited to shapes with unchanged boundaries. Conformal maps [Gu et al. 07, Zeng et al. 10] have been applied to register non-rigid surfaces, by flattening a 3D surface to a 2D domain. However, they are usually accompanied by model cutting and hole filling. Therefore, they are very sensitive to topology and boundary changes. In [Ahmed et al. 08], barycentric coordinates of harmonic functions were utilized to establish dense correspondences between shape-from-silhouette surfaces. Multi-dimensional scaling has also been employed to register non-rigid surfaces. Bronstein et al. [Bronstein et al. 06] proposed the generalized MDS to embed 2-manifolds and then match them. Jain and Zhang [Jain and Zhang 06] proposed a shape matching framework by spectral embedding of the geodesic affinity matrices and thin-plate splines. It requires computing geodesics between each pair of points, and decomposing a dense affinity matrix. For incomplete shapes, Tevs et al. [Tevs et al. 09] used the RANSAC-like algorithm to match points and refine the registration by the post process based on geodesics. Although it does not enforce strict requirements on topology and boundary of the shape, the geodesics are sensitive to noise and holes.

To establish a compact matching, dense registration typically needs some intrinsic coordinates to parameterize and index points on the shape. Since we are concentrating on partial shapes of natural nonrigid objects (e.g., faces, articulated objects), the coordinates must be invariant to natural deformation, and they should be determined only by local geometry while avoiding negative effects from changing boundaries and topological variation. Heat kernel, as the fundamental solution of the heat diffusion on manifolds, has been applied to indexing points in the heat kernel map [Ovsjanikov et al. 10] and the heat triangulation [Jones et al. 08]. It measures the heat transferred from one point to another. As time increases, heat spreads out towards a growing neighborhood that elegantly bridges the local and global characteristics in a multiscale sense. This multiscale property gives rise to an intrinsic connection between the diffusion and time. Heat kernel is also relevant to the statistical probability of connecting paths (e.g., random walk in a Brownian motion). Therefore, it is very stable under inelastic deformation, noise, and small topological holes, which is destined as a promising tool for shape registration as well as other problems in computer graphics.

12.2.2 Intrinsic Coordinates

In [Ovsjanikov et al. 10], the heat kernel map (HKM) was defined by the following map

$$\Phi_p^M : M \to F, \qquad \Phi_p^M = h_t^M(p, x), \tag{12.13}$$

where p is a fixed source point, and F is the space of functions from $\mathbb{R}^+$ to $\mathbb{R}^+$. Thus, Φ_p^M associates with every point $x \in M$ a real-valued function of one parameter t given by $h_t^M(p, x)$. The source point p may be any point on manifold M. For a generic connected compact manifold M without boundary and a generic point p on M, the HKM is injective, which indicates $\Phi_p^M(x) = \Phi_p^M(y)$ if and only if $x = y$. This result has been proved by [Ovsjanikov et al. 10].

In [Jones et al. 08], the *heat triangulation theorem* was introduced for local parameterization.

Theorem 12.1 (Heat Triangulation Theorem [Jones et al. 08]) *Let M be a smooth, d-dimensional compact manifold, and z be a point on M. Let $B_r(z)$ be an embedded ball with center z and radius r. Let $p_1, ..., p_d$ be d linearly independent directions. Let y_i be so that $y_i - z$ is in the direction p_i, with $c_4 r \leq d_g(y_i, z) \leq c_5 r$ for each $i = 1, ..., d$ and let $t = c_6 r^2$. The map*

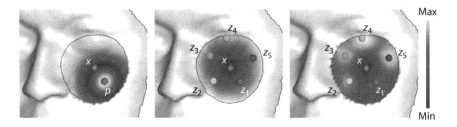

Figure 12.5. Positioning of different maps (from left to right): one-source multi-time HKM, multi-source one-time HKC, and multi-source multi-time HKC.

$\Phi : B_{c_1 r}(z) \to \mathbb{R}^d$, *defined by*

$$x \to (r^d h_t(x, y_1), ..., r^d h_t(x, y_d)),$$

at some t satisfying, for any $x_1, x_2 \in B_{c_1 r}(z)$

$$\frac{c_2}{r} d_g(x_1, x_2) \leq ||\Phi(x_1) - \Phi(x_2)|| \leq \frac{c_3}{r} d_g(x_1, x_2),$$

with constants $c_1, c_2, c_3, c_4, c_5, c_6 > 0$.

This theorem induces a one-to-one map from a local geodesic ball $B_{c_1 r}(z)$ on the manifold to $\mathbb{R}^d$, which therefore proves the injection of heat kernel coordinates in this local ball. However, it needs d points with independent directions from center point z, leading to another problem to solve.

Similarly in [Hou and Qin 12b], heat kernels from multiple sources were used for dense shape registration. For d-dimensional manifold M, the heat kernel coordinates (HKC) were defined as a vector of heat kernels with the number of sources $s > d$:

$$\text{HKC}(x) = [h_t(x, z_1), ..., h_t(x, z_s)]^T, \quad x \in M, \qquad (12.14)$$

where $\{z_1, ..., z_s\}$ are source points on M. Multiple sources can offer direction and distance information during indexing, which is analogous to the trilateration technique in navigation and surveying.

Besides some properties inherited from heat kernels, including intrinsic, stable, non-negative, and multiscale, the heat kernel coordinates are flexible to conditions of geometry and time. Even for a plane with simple geometry, the heat kernel coordinates can position points. Fig. 12.5 shows positioning of different maps: one-source multi-time HKM, multi-source one-time HKC, and multi-source multi-time HKC, where $t \in [4, 1024]$ for multi-time maps. Within the circles, the color of an arbitrary point y denotes $||\text{HKC}(x) -$

HKC$(y)\|_2$. The one-source HKM has similar values in a dark ring around the source point, resulting in ambiguities in location for dense registration. Multi-source HKC restrict similar values of HKC(x) to the dark area close to x, which localizes x with small deviation. Since one-time and multi-time HKC have similar performance of positioning, we will use one-time HKC in our experiments to reduce computation time and tolerate strict time range. Moreover, we use stable matches of features as source points, resulting in feature-driven HKC.

12.2.3 Priority-Vicinity Search

A key problem of registration is searching/matching corresponding points. An available method in existing work is the nearest-neighbor algorithm, which finds the correspondence of a point as its nearest neighbor in the parametric domain. However, it does not consider the geometric compatibility, by which we mean if two points are close to each other in the reference shape, their correspondences in the target shape should also be close. Consequently, there could easily be flips from generated correspondences due to computation error and non-isometric distortion. Furthermore, the global search is time-consuming, as a result, the approximated algorithm is often used instead.

To overcome the aforementioned difficulty, [Hou and Qin 12b] proposed an idea to search correspondences locally in the vicinities of registered points, and propagate correspondence from matched features to other points. That is, for a registered pair (x_1, x_2), a direct neighbor of x_1 only finds its correspondence in the vicinity of x_2. This enforced constraint ensures the geometric compatibility to avoid large flips, but might be trapped into local optima and stop the propagation. It is easy to notice that dense registration has different reliability over the entire shape domain. Features are apparently more reliable than rural plain areas. If the registration starts from unreliable areas, it is possible that mismatch will also accumulate during the propagation. Therefore, the idea is to define a priority for each point, and initiate the vicinity search by choosing the candidate point with the highest priority. Intuitively speaking, greater heat kernels are more reliable than smaller ones. In [Hou and Qin 12b], the magnitudes of heat kernel coordinates were utilized to measure the priority of a point in registration

$$P_t(x) = \|(h_t(x, z_1), \ldots, h_t(x, z_s))\|_2. \qquad (12.15)$$

Given a priority function that somehow measures approximate reliability of registration, active registered points are inserted into a heap. This does not need to be a precise estimation, as the example shown in [Hou and Qin 12b], since it only provides a rough order of the registration process.

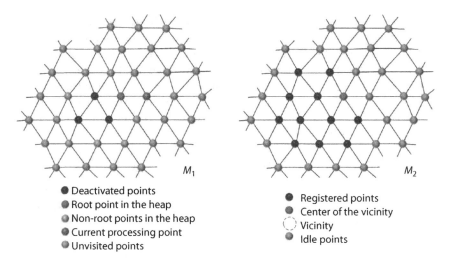

Figure 12.6. Illustration of the vicinity search.

Active registered points are inserted into a heap. Each time, the active point with the greatest priority (the root of the heap) is processed, and vicinity searches are initiated for its 1-ring neighbors. Fig. 12.6 illustrates the procedure of a vicinity search. At this moment, the red point in M_1 is at the root of the heap with the highest priority (key). The blue point is a direct neighbor. The algorithm now finds its corresponding point in the blue-shaded vicinity in M_2. Newly registered points are inserted into the heap, and the current processing point is deactivated. The overall algorithm in Algorithm 11 was named "priority-vicinity search" by [Hou and Qin 12b].

Assuming the vicinity size is bounded by some constant, the computational complexity of this algorithm is $O(V \log H)$, where V is the number of total points, and H is the heap size. In practice, the heap contains the propagation front of registration, which is much smaller than the entire domain. The size of the vicinity, usually set of 2-ring or 3-ring in our experiments, balances the local geometric compatibility and global optimization. That is, large vicinity size increases the global optimization but decreases the geometric compatibility, and vice versa.

As shown in Fig. 12.7, the registration method in [Hou and Qin 12b] proceeds in three main steps: feature detection, feature matching, and dense registration, all of which are hinged upon heat kernels approximated from the Laplace-Beltrami eigenfunctions. Fig. 12.8 shows some registration results from [Hou and Qin 12b], using the above method. Because of the

---

Algorithm 11: Priority-vicinity search.

Input: mesh M_1, M_2, source-pair set S
Output: correspondence set C

Initialize heap $H = \emptyset$;
for *pair* $i = (i_1, i_2)$ *in* S **do**
 insert i_1 with its priority $P_t(i_1)$ into H;
end
while $H \neq \emptyset$ **do**
 x_1 : the root of H;
 delete the root of H;
 for $y_1 \in N(x_1)$ **do**
  if y_1 *is registered* **then**
   continue;
  end
  if $P_t(y_1) < \epsilon_p$ **then**
   continue;
  end
  Initialize y_2;
  Initialize d_{min};
  for $z_2 \in N(x_2)$ **do**
   if $\|\text{HKC}(y_1) - \text{HKC}(z_2)\| < d_{min}$ **then**
    $y_2 = z_2$;
    $d_{min} = \|\text{HKC}(y_1) - \text{HKC}(z_2)\|$;
   end
  end
  Mark y_1 as registered;
  insert (y_1, y_2) to C;
  insert y_1 with its priority $P_t(y_1)$ into H;
 end
end

---

diffusion nature, the uncertainty gradually increases when the heat diffusion approaches to boundaries from sources. As a result, the registration close to the boundary tends to be less trustworthy. Therefore, registration process is stopped at regions near the boundaries (as far as dense registration is concerned).

Moreover, this method can handle some similar but different objects. Fig. 12.9 shows an experimental result, where a woman's face is registered to a man's face. We accept more features in feature detection, aiming to create more candidates for feature matching. We reset the parameter $\sigma_2 =$

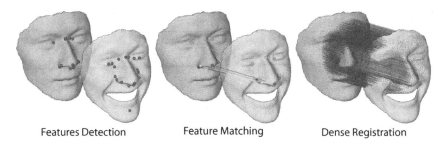

Features Detection   Feature Matching   Dense Registration

Figure 12.7. The architecture highlights a complete solution for shape registration. It computes the Laplace-Beltrami eigenfunctions, by which three steps proceed: feature detection, feature matching, and dense registration.

1 in Eq. (12.6) to increase the affinities of feature pairs, which weighs more on geometric compatibility than feature similarity. This experiment demonstrates that our method has great tolerance on distortion and noise, with a potential to be used for expression or motion transfer. It shows that the nature of our method relies more on the probability and geometric compatibility, not on single-point similarity.

12.2.4   Refined Hierarchical Registration

The above registration method uses a one-level procedure. The registration results can be improved by extending the one-level procedure to a multi-level hierarchical one, as presented in [Zhong et al. 12].

The idea is to generate correspondences in multiple levels in a coarse-to-fine manner, with additional features incrementally inserted in each level. The registration starts from the coarsest resolution. The registration results obtained in one level serve as references for the registration in the next level. The main steps of this method include: (1) Detect and match features to get a small initial set of feature matches; (2) Construct hierarchical structures of input shapes; (3) Perform registration at the coarsest level using the initial feature set; (4) Select some newly registered points as additional features; (5) Perform registration at the next level using results from the previous level and the expanded set of feature references; (6) Repeat steps (4) and (5) until all valid points are registered.

The rationale of the approach is that distinguishing elements that are distant from each other on the surface are much more accurate than nearby elements. Even with a small number of features, we can achieve very good registration on a heavily downsampled version of the original shapes. The

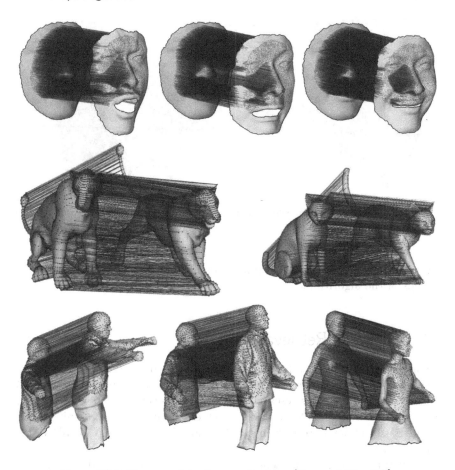

Figure 12.8. Shape registration results from [Hou and Qin 12b].

registration result of a coarse resolution can serve as seed correspondences when performing registration in a finer level. The large number of available seeds significantly reduces the chances of correspondences being trapped in an incorrect location. Moreover, the multi-resolution process enables us to pick additional features from already registered points.

We then select from R_l some vertex pairs as new features and insert them into the feature set. These new added feature pairs should be both reliable (having great matching score) and not in the δ-neighborhood of any existing feature points. The expanded feature set C_l enables more discriminative heat kernel coordinates in the next level. We carry on this process from the coarsest level to the finest level until we obtain the final registration

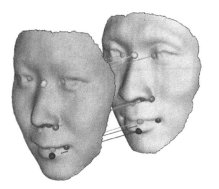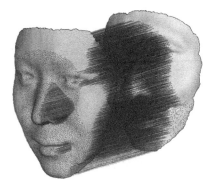

Figure 12.9. Registration result on similar but different shapes. A woman's face is registered to a man's face.

set R_0 between the original meshes S_0 and T_0. Fig. 12.10 shows the major steps of the algorithm.

12.3   Shape Retrieval

For a given database and a query shape, the problem of retrieval is to find similar shapes from the database. It is a high-level problem that involves many techniques. This research is still in its developing stage. Some representative works include the 3D search engine by Princeton University [1], and newly exposed results in the Eurographics workshop on 3D object retrieval [2].

In this section, we present a shape retrieval work based on heat kernels and features, to glimpse this research direction. The core problem is simplified representation of 3D shapes possibly with deformation.

12.3.1   Related Work

The main task of shape retrieval is to find effective shape descriptors for similarity measure. One direct approach is to project the 3D model to 2D planes at different views [Chen et al. 03]. Zernike moments and Fourier descriptors of the projected silhouettes were adopted as representations for retrieval. The similar idea can be found in [Lee et al. 06], where the aspect graph representation was employed as a shape descriptor.

---

[1] http://shape.cs.princeton.edu/search.html
[2] http://eg3dor2012.ist.utl.pt/, http://3dor2013.di.univr.it/, http://3dor2014.ensea.fr/

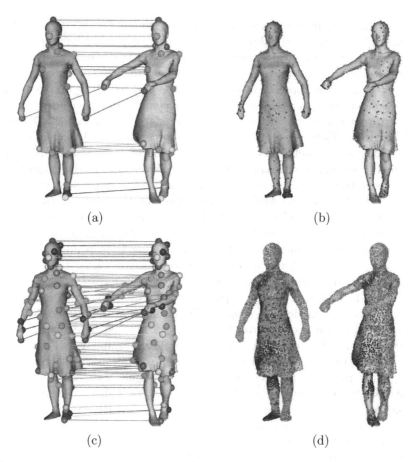

(a)

(b)

(c)

(d)

Figure 12.10. Major steps in the hierarchical registration algorithm. The blue shape is the source and the red one is the target. We use a three level hierarchy in this example. (a) Initial feature correspondences; (b) Coarse registration result (Third level); (c) Expanded feature correspondences (Third level); (d) Final registration result. See color insert.

Global shape descriptors have been widely used in shape retrieval, because of their efficient computation. In [Osada et al. 02], the shape distribution was introduced based on the measurements of distance, angle, and area between random points on the surface. In [Yu et al. 03], geometric and topological feature maps were used for comparing shapes, which capture the amount of effort required to morph a 3D object into a canonical sphere. Spherical harmonics [Funkhouser et al. 03] as basis for shape representation were integrated into a search engine for 3D models. Laplace-spectra [Reuter et al. 05], eigenvalues of the Laplace operator, were also used as fingerprints for shape matching. In [Tam and Lau 07], a method for deformable model retrieval was proposed, using topological and geometric features. The topological features compose a skeleton representation of the model. Three types of geometric information are then computed and associated with the topological features. In [Gal et al. 07], a pose-oblivious shape signature was introduced, which is a combination of the distribution of two scalar functions defined on the boundary surface of the 3D shape. Reeb graph [Tierny et al. 09] has been adopted for partial shape retrieval. Partial similarity between two shapes is evaluated by computing a variant of their maximum common sub-graph. In [Sfikas et al. 11], the graph-based representation was combined with conformal geometry for non-rigid shape retrieval. More related work of shape representation and similarity measurement can be found in a survey [Tangelder and Veltkamp 08].

Recently, bag-of-words (BoW) methods prevail in shape retrieval, which is unsurprisingly coincident with the popular methods in image retrieval. It can be traced back to the previous work of shape topics [Liu et al. 06], where the spin-images [Johnson and Hebert 99] are used as a shape descriptor. In [Toldo et al. 10], a part-based representation was utilized by partitioning the model into subparts. In [Tabia et al. 10], a descriptor was integrated into a BoW approach, which is an indexed collection of closed curves on the 3D surface. In [Lavoué 11], uniform sampling and local spectral descriptor were adopted for partial shape retrieval. The shape google, originally proposed by Ovsjanikov et al. [Ovsjanikov et al. 09], employs the heat kernel signature (HKS) [Sun et al. 09] as a shape descriptor, and computes frequencies of words in a vocabulary. The HKS is a concise and informative representation, which preserves all information about the intrinsic geometry of the shape. Later, a scale-invariant heat kernel signature (SI-HKS) [Bronstein and Kokkinos 10] was proposed to solve scale changes for this approach. In [Bronstein et al. 11, Ovsjanikov et al. 09], the shape google also introduced the spatially-sensitive bag-of-words (SS-BoW) by looking at frequencies of word pairs, with encoded spatial relations.

The challenges of shape retrieval hinge upon representation paradigms. An intrinsic paradigm of data representation is a graph connecting all the points. A shape can be faithfully represented by its affinity matrices formed by local or global structures, such as Laplacian matrix, geodesic matrix, diffusion matrix, and heat kernel matrix. Applying the graph paradigm to shape comparison oftentimes requires solving the eigen-decomposition problem of affinity matrices, and projecting the shape to its spectral basis [Reuter et al. 06] or principle directions [Bronstein et al. 06]. Yet, the graph could be dominated by the majority of non-salient points. This deteriorates its discrimination power in shape retrieval. The shape google employs the bag-of-words, or called bag-of-features in some literature. It categorizes the points to different geometric words from a given vocabulary. Frequencies of geometric words spreading on the entire shape form a concise shape descriptor. Word frequency is a good descriptor for text data. It may not be suitable for images and shapes, where spatial information is particularly important. Although the shape google introduces a spatially-sensitive bag-of-words, it excessively suppresses the information by counting frequencies of the word pairs. Besides, the SS-BoW severely increases the time complexity by computing point-to-point heat kernels.

In [Hou et al. 12a], a new paradigm, called bag-of-feature-graphs (BoFG), was proposed. It was motivated by the urgent need for a concise and spatially-informative representation for shape comparison and retrieval. The key idea is to construct graphs of features on the shape. Given a vocabulary of geometric words, corresponding to each word we build a graph that records spatial information between features, weighted by their similarities to this word.

12.3.2 Shape Google

The shape google method is introduced by [Ovsjanikov et al. 09]. It utilizes a bag-of-words (BoW) scheme based on the heat kernel signature $\text{HKS}(x)$ as shape representation. Let $\{W_1, \dots, W_V\}$ be a vocabulary of geometric words with size V. The words $\{W_i\}$ are representation vectors in the descriptor space clustered by the k-means algorithm. For each point x, the shape google computes its word distribution

$$\Theta(x) = [\theta_1(x), \dots, \theta_V(x)]^T.$$

The similarity of x and word W_i is given by

$$\theta_i(x) = c(x)e^{-\frac{\|\text{HKS}(x) - W_i\|^2}{2\sigma^2}}, \tag{12.16}$$

where σ is a parameter, and $c(x)$ is a constant for normalization. The BoW descriptor of a surface M is computed by integrating word similarities over the entire shape

$$f(M) = \int_M \Theta(x)d\mu(x), \tag{12.17}$$

where $\mu(x)$ denotes the surface area of x. As shown in Fig. 12.11, the BoW descriptor is a $V \times 1$ vector that measures the frequencies of words appearing on the shape. The shape google also introduced a scale-invariant bag-of-words (SS-BoW) descriptor, given by

$$F(M) = \int_{M \times M} \Theta(x)\Theta^T(y)h_t(x, y)d\mu(x)d\mu(y). \tag{12.18}$$

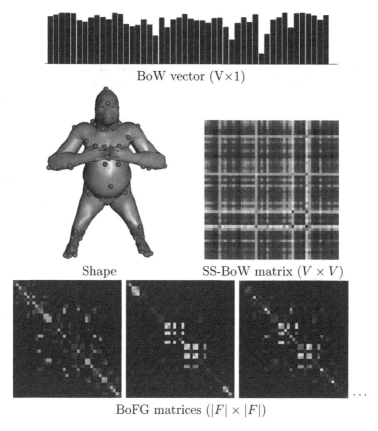

BoW vector (V×1)

Shape   SS-BoW matrix $(V \times V)$

BoFG matrices $(|F| \times |F|)$

Figure 12.11. Shape descriptors. The BoW descriptor is a V×1 vector of word frequencies on the shape. The SS-BoW descriptor is a V×V matrix of word-pair frequencies. The BoFG contains a series of |F|×|F| matrices that characterize spatial information of features clustered into different word categories. The near-zero entries in a BoFG matrix indicate they are hardly classified to this category, and therefore, not considered in this graph.

As shown in Fig. 12.11, it is a $V \times V$ matrix that measures frequencies of word pairs.

The BoW descriptor is a $V \times 1$ vector of word frequencies on the shape. The SS-BoW descriptor is a $V \times V$ matrix of word-pair frequencies. The BoFG contains a series of $|F| \times |F|$ matrices that characterize spatial information of features clustered into different word categories. The near-zero entries in a BoFG matrix indicate they are hardly classified to this category, and therefore, not considered in this graph.

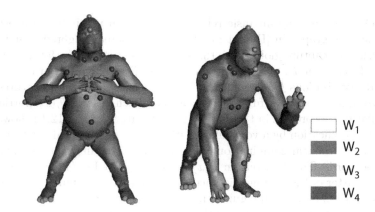

Figure 12.12. Fuzzy classification of features by a vocabulary with four geometric words. A feature is colored by a linear combination of word-colors, according to its similarities of the words.

According to the Informative Theorem in [Sun et al. 09], the heat kernel signature contains all the information of heat kernels. Thus, the SS-BoW has no more geometry information than the BoW before integration. For the ease of comparison, the shape google highly suppresses the geometry information by computing the frequencies of words or word-pairs on the shape. It ends up with concise descriptors for comparison, yet completely loses spatial information. Besides, the shape-google algorithms are time-consuming, since they are working on all the points in the data. The BoW needs computing HKS values of all points, while the SS-BoW needs computing all point-to-point heat kernels. Assume the time complexity for computing a HKS descriptor is $O(D)$. For a shape with N points, the time complexity of BoW is $O(ND)$, and SS-BoW is $O(N^2D)$ which is quadratic to N.

12.3.3   Bag of Feature Graphs

To reduce the complexity of the shape google, one needs to reduce the number of points involved in representing the shape. A straightforward solution is to select feature points, which keep most information of the shape geometry. Multi-scale features contain geometry information ranging from points in small scales to the entire shape in large scales. Instead of counting word frequencies, the method in [Hou et al. 12a] constructs graphs on detected features, giving rise to a bag-of-feature-graphs (BoFG) scheme. The graphs encode spatial relations between features, which contain much more geometry information in representing the shape.

One can use any intrinsic relation function on manifolds to construct the feature graph. In [Hou et al. 12a], they adopted weighted heat kernel matrices to capture global structures of graphs. Specifically, for a shape M with feature set F, only feature points $x \in F$ are involved in computing word distributions $\Theta(x)$, which reduces much computation. Features are vector-quantized by a fuzzy classification, which assigns $\theta_i(x)$ portion of similarity to word W_i in the distribution of feature x. Fig. 12.12 shows the fuzzy classification by a vocabulary with four geometric words. A feature is colored by a linear combination of word-colors, according to its similarities to the words. The distribution $\Theta(x)$ is computed by Eq. (12.16) with σ set as a quarter of the average distance of words in the vocabulary. This fuzzy classification reduces ambiguities in graph comparison, and also avoids mis-classification in a hard quantization. For a geometric word W_i, matrix G_i is constructed by entries $G_i(x, y)$

$$G_i(x, y) = \theta_i(x)\theta_i(y)h_t(x, y). \tag{12.19}$$

It is the heat kernel between x and y weighted by their similarities to the geometric word W_i.

The matrix set $\mathbf{G}(M) = \{G_1, \ldots, G_V\}$ comprises a BoFG representation of the shape M. As shown in the bottom row of Fig. 12.11, matrices characterize spatial information of features assigned to different word categories. The near-zero entries in a matrix indicate they are hardly classified to this category, and therefore, not considered in this graph. It contains all the geometric information of features in a multi-scale way, which faithfully characterizes the shape. The computation complexity for this matrix representation is $O(|F|^2 D)$, as the computed heat kernels can be shared by all matrices. Considering the size of feature set is always much less than the total number of points on the shape, the BoFG is much faster than the shape google.

12.3.4   Shape retrieval

The BoFG descriptor consists of leading eigenvalues of BoFG matrices. Each G_i is a real symmetric matrix, whose eigenvalues are all real and eigenvectors are perpendicular to each other. In [Hou et al. 12a], six leading eigenvalues of a matrix were used to form $S_i(M)$. A $6V \times 1$ vector $[S_1(M), \ldots, S_V(M)]^T$ was adopted as a concise shape descriptor. This reduces the dimension of the matrix by multi-dimensional scaling [Bronstein et al. 06]. Fig. 12.13 shows some non-rigid shapes and their BoFG descriptors. The deformed cat-models have very similar BoFG descriptors, while the horse-model has a quite different one. It projects the matrix to its main directions with coordinates leaving in $S_i(M)$, which are stable to a small amount of outliers.

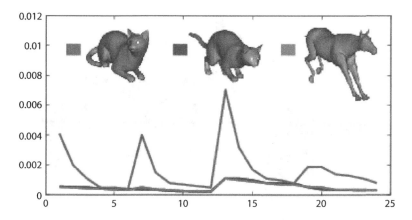

Figure 12.13. Some non-rigid shapes and their BoFG descriptors. The cat-models have very similar descriptors, different with the horse-model.

The similarity distance between two shapes M_1 and M_2 is defined as

$$d(M_1, M_2) = \sum_{i=1}^{V} \|S_i(M_1) - S_i(M_2)\|_2. \qquad (12.20)$$

The above distance is based on one-scale heat kernels, which can be easily extended to multi-scale by averaging distances of heat kernels at different values of t.

The BoFG enables partial shape retrieval, where the query shape is only a part of a complete model. The primary challenge here is that partial shapes have different word-frequencies from complete ones. Consequently, the BoW is not suitable to handle partial shape retrieval. In BoFG, a partial shape is represented by a sub-graph of the complete feature graph. It then only needs to extract corresponding sub-graphs from the complete ones. This can be solved by graph matching that is resilient in combating outliers. In [Hou et al. 12a], the spectral matching was employed for partial shapes. It uses a second-order graph matching to match them. The graph matching is resilient in combating outliers, and therefore, is superior for partial shapes. Fig. 12.14 shows the graph matching for partial shape retrieval.

Fig. 12.15 shows some examples of shape retrieval by this method, where top-six retrieved shapes are listed for each query. In the bottom row, models of horse and centaur are retrieved for a query of a partial shape cut from a centaur model.

As a new and powerful paradigm for shape representation, the BoFG offers a concise and faithful representation for shape comparison and

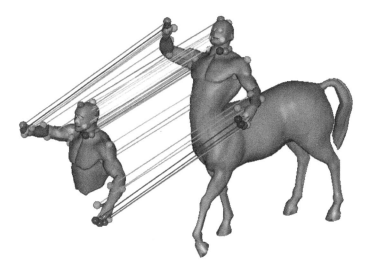

Figure 12.14. Graph matching for partial shape retrieval. Corresponding sub-feature-graphs are extracted to construct new BoFG descriptors for shape comparison.

retrieval. Its manifestation expediently equips with other techniques, such as graph matching, multi-dimensional scaling, and graph spectrum. In such a way, the BoFG is capable of handling partial shape retrieval properly. Because of the feature-based property, its performance heavily depends on feature detection. Therefore, it should be employed in conjunction with stable and multi-scale methods for feature extraction on deformable shapes (e.g., heat kernel signature and wavelet signature).

Figure 12.15. Some example query shapes (Left) and their top-six retrieved shapes (Right). In the bottom row, models of horse and centaur are retrieved for a query of a partial shape cut from a centaur model.

Bibliography

[Ahmed et al. 08] N. Ahmed, C. Theobalt, C. Rössl, S. Thrun, and H.-P. Seidel. "Dense correspondence finding for parametrization-free animation reconstruction from video." In *IEEE Conference on Computer Vision and Pattern Recognition*, pp. 1–8, 2008.

[Alexa 03] M. Alexa. "Differential coordinates for local mesh morphing and deformation." *The Visual Computer* 19:2-3 (2003), 105–114.

[Antoine and Vandergheynst 99] J.-P. Antoine and Pierre Vandergheynst. "Wavelets on the 2-sphere: A group-theoretical approach." *Applied and Computational Harmonic Analysis* 7:3 (1999), 262–291.

[Antoine et al. 10] J.-P. Antoine, D. Roşca, and P. Vandergheynst. "Wavelet transform on manifolds: Old and new approaches." *Applied and Computational Harmonic Analysis* 28:2 (2010), 189–202.

[Arya et al. 98] S. Arya, D. M. Mount, N. S. Netanyahu, R. Silverman, and A. Y. Wu. "An Optimal Algorithm for Approximate Nearest Neighbor Searching in Fixed Dimensions." *Journal of the ACM* 45:6 (1998), 891–923.

[Aubry et al. 11] M. Aubry, U. Schlickewei, and D. Cremers. "The Wave Kernel Signature: A Quantum Mechanical Approach to Shape Analysis." In *International Conference on Computer Vision Workshops*, pp. 1626–1633, 2011.

[Belkin et al. 08] M. Belkin, J. Sun, and Y. Wang. "Discrete laplace operator on meshed surfaces." In *ACM Symposium on Computational Geometry*, pp. 278–287, 2008.

[Belkin et al. 09] M. Belkin, J. Sun, and Y. Wang. "Constructing Laplace operator from point clouds in R^d." In *ACM-SIAM Symposium on Discrete Algorithms*, pp. 1031–1040, 2009.

[Ben-Chen and Gotsman 05] M. Ben-Chen and C. Gotsman. "On the optimality of spectral compression of mesh data." *ACM Transactions on Graphics* 24:1 (2005), 60–80.

[Bertram et al. 00] M. Bertram, M. A. Duchaineau, B. Hamann, and K. Joy. "Bicubic subdivision-surface wavelets for large-scale isosurface representation and visualization." In *IEEE Visualization*, pp. 389–396, 2000.

[Bertram et al. 04] M. Bertram, M. A. Duchaineau, B. Hamann, and K. I. Joy. "Generalized b-spline subdivision-surface wavelets for geometry compression." *IEEE Transactions on Visualization and Computer Graphics* 10:3 (2004), 326–338.

[Bigun et al. 91] J. Bigun, G. H. Granlund, and J. Wiklund. "Multidimensional Orientation Estimation with Applications to Texture Analysis and Optical Flow." *IEEE Transactions on Pattern Analysis and Machine Intelligence* 13:8 (1991), 775–790.

[Boggess and Raich 10] A. Boggess and A. Raich. "Heat Kernels, Smoothness Estimates and Exponential Decay." *ArXiv e-prints*.

[Bonneau 99] G.-P. Bonneau. "Optimal triangular Harr bases for spherical data." In *IEEE Visualization*, pp. 279–284, 1999.

[Bronstein and Bronstein 11] M. M. Bronstein and A. M. Bronstein. "Shape recognition with spectral distances." *IEEE Transactions on Pattern Analysis and Machine Intelligence* 33:5 (2011), 1065–1071.

[Bronstein and Kokkinos 10] M. M. Bronstein and I Kokkinos. "Scale-invariant heat kernel signatures for non-rigid shape recognition." In *IEEE Conference on Computer Vision and Pattern Recognition*, pp. 1704–1711, 2010.

[Bronstein et al. 06] A. M. Bronstein, M. M. Bronstein, and R. Kimmel. "Generalized multidimensional scaling: A framework for isometry-invariant partial surface matching." *Proceedings of the National Academy of Sciences* 103:5 (2006), 1168–1172.

[Bronstein et al. 10a] A. M. Bronstein, M. M. Bronstein, U. Castellani, B. Falcidieno, A. Fusiello, A. A. Godil, L. J. Guibas, I. Kokkinos, Z. Lian, M. Ovsjanikov, G. Patane, M. Spagnuolo, and R. Toldo. "SHREC 2010: robust large-scale shape retrieval benchmark." In *Eurographics Workshop on 3D Object Retrieval*, 2010.

[Bronstein et al. 10b] A. M. Bronstein, M. M. Bronstein, M. Mahmoudi, R. Kimmel, and G. Sapiro. "A Gromov-Hausdorff framework with diffusion geometry for topological-robust non-rigid shape matching." *International Journal of Computer Vision* 89:2-3 (2010), 266–286.

[Bronstein et al. 11] A. M. Bronstein, M. M. Bronstein, L. J. Guibas, and M. Ovsjanikov. "Shape Google: Geometric Words and Expressions for Invariant Shape Retrieval." *ACM Transactions on Graphics* 30:1 (2011), 1:1–1:20.

[Castellani et al. 08] U. Castellani, M. Cristani, S. Fantoni, and V. Murino. "Sparse points matching by combining 3D mesh saliency with statistical descriptors." *Computer Graphics Forum* 27:2 (2008), 643–652.

[Catmull and Clark 78] E. Catmull and J. Clark. "Recursively generated B-spline surfaces on arbitrary topological meshes." *Computer-Aided Design* 10:6 (1978), 350–365.

[Chan and Shen 05] T. Chan and J. Shen. *Image Processing And Analysis: Variational, Pde, Wavelet, And Stochastic Methods*. Society for Industrial and Applied Mathematics, 2005.

[Chang and Zwicker 08] W. Chang and M. Zwicker. "Automatic registration for articulated shapes." *Computer Graphics Forum* 27:5 (2008), 1459–1468.

[Charina et al. 10] M. Charina, C. K. Chui, and W. He. "Tight frames of compactly supported multivariate multi-wavelets." *Journal of Computational and Applied Mathematics* 233:8 (2010), 2044–2061.

[Chavel 84] I. Chavel. *Eigenvalues in Riemannian Geometry.* Academic Press, 1984.

[Chen et al. 03] D.-Y. Chen, X.-P. Tian, Y.-T. Shen, and M. Ouhyoung. "On Visual Similarity Based 3D Model Retrieval." *Computer Graphics Forum* 22:3 (2003), 223–232.

[Christensen 02] O. Christensen. *An Introduction to Frames and Riesz Bases.* Birkhäuser Boston, 2002.

[Chuang and Kazhdan 11] M. Chuang and M. Kazhdan. "Interactive and Anisotropic Geometry Processing Using the Screened Poisson Equation." *ACM Transactions on Graphics* 30:4 (2011), 57:1–57:10.

[Clarenz et al. 00] U. Clarenz, U. Diewald, and M. Rumpf. "Anisotropic Geometric Diffusion in Surface Processing." In *IEEE Visualization*, pp. 397–405, 2000.

[Coifman and Lafon 06] R. R. Coifman and S. Lafon. "Diffusion maps." *Applied and Computational Harmonic Analysis* 21:1 (2006), 5–30.

[Coifman and Maggioni 06] R. R. Coifman and M. Maggioni. "Diffusion wavelets." *Applied and Computational Harmonic Analysis* 21:1 (2006), 53–94.

[Cour et al. 06] T. Cour, P. Srinivasan, and J. Shi. "Balanced graph matching." In *Neural Information Processing Systems*, pp. 313–320, 2006.

[Craciun et al. 05] G. Craciun, M. Jiang, D. Thompson, and R. Machiraju. "Spatial Domain Wavelet Design for Feature Preservation in Computational Data Sets." *IEEE Transactions on Visualization and Computer Graphics* 11:2 (2005), 149–159.

[Daubechies et al. 99] I. Daubechies, I. Guskov, P. Schröder, and W. Sweldens. "Wavelets on Irregular Point Sets." *Philosophical Transactions of the Royal Society A* 357:1760 (1999), 2397–2413.

[Daubechies 92] I. Daubechies. *Ten Lectures on Wavelets.* CBMS-NSF Regional Conference Series in Applied Mathematics, Society for Industrial and Applied Mathematics, 1992.

[de Iehl and Péroche 00] J. C. de Iehl and B. Péroche. "An Adaptive Spectral Rendering with a Perceptual Control." *Computer Graphics Forum* 19:3 (2000), 291–300.

[Desbrun et al. 00] M. Desbrun, M. Meyer, P. Schröder, and A. H. Barr. "Anisotropic Feature-Preserving Denoising of Height Fields and Bivariate Data." In *Graphics Interface*, pp. 145–152, 2000.

[Dey et al. 10] T. K. Dey, K. Li, C. Luo, P. Ranjan, I. Safa, and Y. Wang. "Persistent heat signature for pose-oblivious matching of incomplete models." *Computer Graphics Forum* 29:5 (2010), 1545–1554.

[Dubrovina and Kimmel 10] A. Dubrovina and R. Kimmel. "Matching shapes by eigendecomposition of the Laplace-Beltrami operator." In *3D Data Processing, Visualization and Transmission*, 2010.

[Duchenne et al. 09] O. Duchenne, F. Bach, I. Kweon, and J. Ponce. "A tensor-based algorithm for high-order graph matching." In *IEEE Conference on Computer Vision and Pattern Recognition*, pp. 1980–1987, 2009.

[Fleishman et al. 03] S. Fleishman, I. Drori, and D. Cohen-Or. "Bilateral Mesh Denoising." *ACM Transactions on Graphics* 22:3 (2003), 950–953.

[Fouss et al. 07] F. Fouss, A. Pirotte, J.-M. Renders, and M. Saerens. "Random-Walk Computation of Similarities between Nodes of a Graph with Application to Collaborative Recommendation." *IEEE Transactions on Knowledge and Data Engineering* 19:3 (2007), 355–369.

[Freeden and Windheuser 97] W. Freeden and U. Windheuser. "Combined spherical harmonic and wavelet expansion - A future concept in Earth's gravitational determination." *Applied and Computational Harmonic Analysis* 4:1 (1997), 1–37.

[Funkhouser et al. 03] T. Funkhouser, P. Min, M. Kazhdan, J. Chen, A. Halderman, D. Dobkin, and D. Jacobs. "A Search Engine for 3D Models." *ACM Transactions on Graphics* 22:1 (2003), 83–105.

[Gal et al. 07] R. Gal, A. Shamir, and Daniel Cohen-Or. "Pose-Oblivious Shape Signature." *IEEE Transactions on Visualization and Computer Graphics* 13:2 (2007), 261–271.

[Grigor'yan 99] A. Grigor'yan. "Escape rate of Brownian motion on Riemannian manifolds." *Applicable Analysis* 71:1 (1999), 63–89.

[Grigor'yan 06] A. Grigor'yan. "Heat kernels on weighted manifolds and applications." *Contemporary Mathematics* 398 (2006), 93–191.

[Gu et al. 07] X. Gu, S. Wang, J. Kim, Y. Zeng, Y. Wang, H. Qin, and D. Samaras. "Ricci flow for 3D shape analysis." In *International Conference on Computer Vision*, 2007.

[Guskov et al. 99] Igor Guskov, Wim Sweldens, and Peter Schröder. "Multiresolution signal processing for meshes." In *ACM SIGGRAPH*, pp. 325–334, 1999.

[Hammond et al. 11] D. K. Hammond, P. Vandergheynst, and R. Gribonval. "Wavelets on graphs via spectral graph theory." *Applied and Computational Harmonic Analysis* 30:2 (2011), 129–150.

[Han et al. 09] S. Han, W. Tao, D. Wang, X. C. Tai, and X. Wu. "Image Segmentation Based on GrabCut Framework Integrating Multiscale Nonlinear Structure Tensor." *IEEE Transactions on Image Processing* 18:10 (2009), 2289–2302.

[Hastie and Tibshirani 96] T. Hastie and R. Tibshirani. "Discriminant Adaptive Nearest Neighbor Classification." *IEEE Transactions on Pattern Analysis and Machine Intelligence* 18:6 (1996), 607–616.

[Hildebrandt and Polthier 04] K. Hildebrandt and K. Polthier. "Anisotropic Filtering of Non-Linear Surface Features." *Computer Graphics Forum* 23:3 (2004), 391–400.

[Holschneider 96] M. Holschneider. "Continuous wavelet transforms on the sphere." *Journal of Mathematical Physics* 37 (1996), 4156–4165.

[Horn and Johnson 85] R. A. Horn and C. R. Johnson. *Matrix Analysis.* Cambridge University Press, 1985.

[Hou and Qin 10] T. Hou and H. Qin. "Efficient computation of scale-space features for deformable shape correspondences." In *European Conference on Computer Vision,* pp. 384–397, 2010.

[Hou and Qin 12a] T. Hou and H. Qin. "Continuous and Discrete Mexican Hat Wavelet Transforms on Manifolds." *Graphical Models* 74:4 (2012), 221–232.

[Hou and Qin 12b] T. Hou and H. Qin. "Robust Dense Registration of Partial Nonrigid Shapes." *IEEE Transactions on Visualization and Computer Graphics* 18:8 (2012), 1268–1280.

[Hou and Qin 13] T. Hou and H. Qin. "Admissible Diffusion Wavelets and Their Applications in Space-Frequency Processing." *IEEE Transactions on Visualization and Computer Graphics* 19:1 (2013), 3–15.

[Hou et al. 12a] T. Hou, X. Hou, M. Zhong, and H. Qin. "Bag-of-Feature-Graphs: A New Paradigm for Non-rigid Shape Retrieval." In *International Conference on Pattern Recognition,* pp. 1513–1516, 2012.

[Hou et al. 12b] T. Hou, M. Zhong, and H. Qin. "Diffusion-Driven High-Order Matching of Partial Deformable Shapes." In *International Conference on Pattern Recognition,* pp. 137–140, 2012.

[Hsu 89] P. Hsu. "Heat Semigroup on A Complete Riemannian Manifold." *The Annals of Probability* 17:3 (1989), 1248–1254.

[Hua et al. 08] J. Hua, Z. Lai, M. Dong, X. Gu, and H. Qin. "Geodesic distance-weighted shape vector image diffusion." *IEEE Transactions on Visualization and Computer Graphics* 14:6 (2008), 1643–1650.

[Huang et al. 08] Q.-X. Huang, B. Adams, M. Wicke, and L. J. Guibas. "Non-rigid registration under isometric deformations." *Computer Graphics Forum* 27:5 (2008), 1449–1457.

[Jaffard et al. 01] S. Jaffard, Y. Meyer, and R. D. Ryan. *Wavelets: Tools for Science & Technology.* Society for Industrial and Applied Mathematics, 2001.

[Jain and Zhang 06] V. Jain and H. Zhang. "Robust 3D shape correspondence in the spectral domain." In *Shape Modeling International,* pp. 19–30, 2006.

[Jain et al. 07] V. Jain, H. Zhang, and O. van Kaick. "Non-rigid spectral correspondence of triangle meshes." *International Journal of Computer Vision* 13:1 (2007), 101–124.

[Jin et al. 05] S. Jin, R. R. Lewis, and D. West. "A Comparison of Algorithms for Vertex Normal Computation." *The Visual Computer* 21:1-2 (2005), 71–82.

[Johnson and Hebert 99] A. E. Johnson and M. Hebert. "Using Spin-images for efficient multiple model recognition in cluttered 3-D scenes." *IEEE Transactions on Pattern Analysis and Machine Intelligence* 21:5 (1999), 433–449.

[Jones et al. 03] T. R. Jones, F. Durand, and M. Desbrun. "Non-Iterative, Feature-Preserving Mesh Smoothing." *ACM Transactions on Graphics* 22:3 (2003), 943–949.

[Jones et al. 08] P. W. Jones, M. Maggioni, and R. Schul. "Manifold parameterizations by eigenfunctions of the Laplacian and heat kernels." *Proceedings of the National Academy of Sciences* 105:6 (2008), 1803–1808.

[Joshi et al. 07] P. Joshi, M. Meyer, T. DeRose, B. Green, and T. Sanocki. "Harmonic coordinates for character articulation." *ACM Transactions on Graphics* 26:3 (2007), 71:1–71:9.

[Ju et al. 05] T. Ju, S. Schaefer, and J. Warren. "Mean value coordinates for closed triangular meshes." *ACM Transactions on Graphics* 24:3 (2005), 561–566.

[Karni and Gotsman 00] Z. Karni and C. Gotsman. "Spectral Compression of Mesh Geometry." In *ACM SIGGRAPH*, pp. 279–286, 2000.

[Kazhdan et al. 02] M. Kazhdan, B. Chazelle, D. Dobkin, A. Finkelstein, and T. Funkhouser. "A Reflective Symmetry Descriptor." In *European Conference on Computer Vision*, pp. 642–656, 2002.

[Kazhdan et al. 03] M. Kazhdan, T. Funkhouser, and S. Rusinkiewicz. "Rotation Invariant Spherical Harmonic Representation of 3D Shape Descriptors." In *Symposium on Geometry Processing*, pp. 156–164, 2003.

[Kazhdan 04] M. Kazhdan. "Shape Representations and Algorithms for 3D Model Retrieval." Ph.D. thesis, Princeton University, 2004.

[Kim and Varshney 06] Y. Kim and A. Varshney. "Saliency-guided Enhancement for Volume Visualization." *IEEE Transactions on Visualization and Computer Graphics* 12:5 (2006), 925–932.

[Kim et al. 09] H. S. Kim, H. K. Choi, and K. H. Lee. "Feature detection of triangular meshes based on tensor voting theory." *Computer-Aided Design* 41:1 (2009), 47–58.

[Kim et al. 10] Y. Kim, A. Varshney, D. W. Jacobs, and F. Guimbretière. "Mesh Saliency and Human Eye Fixations." *ACM Transactions on Applied Perception* 7:2 (2010), 12:1–12:13.

[Kim et al. 13] W. H. Kim, M. K. Chung, and V. Singh. "Multi-resolution Shape Analysis via Non-Euclidean Wavelets: Applications to Mesh Segmentation and Surface Alignment Problems." In *IEEE Conference on Computer Vision and Pattern Recognition*, pp. 2139–2146, 2013.

[Kolluri et al. 04] R. Kolluri, J. R. Shewchuk, and J. F. O'Brien. "Spectral Surface Reconstruction From Noisy Point Clouds." In *Symposium on Geometry Processing*, pp. 11–21, 2004.

[Kraevoy and Sheffer 04] V. Kraevoy and A. Sheffer. "Cross-parameterization and compatible remeshing of 3D models." *ACM Transactions on Graphics* 23:3 (2004), 861–869.

[Lafon et al. 06] S. Lafon, Y. Keller, and R. R. Coifman. "Data Fusion and Multicue Data Matching by Diffusion Maps." *IEEE Transactions on Pattern Analysis and Machine Intelligence* 28:11 (2006), 1784–1797.

[Lavoué 11] G. Lavoué. "Bag of Words and Local Spectral Descriptor for 3D Partial Shape Retrieval." In *Eurographics Workshop on 3D Object Retrieval*, pp. 41–48, 2011.

[Lee et al. 05] C. H. Lee, A. Varshney, and D. W. Jacobs. "Mesh saliency." *ACM Transactions on Graphics* 24:3 (2005), 659–666.

[Lee et al. 06] S. Lee, S. Yoon, I. D. Yun, D. H. Kim, K. M. Lee, and S. U. Lee. "A New 3-D Model Retrieval System Based on Aspect-Transition Descriptor." In *European Conference on Computer Vision*, pp. 543–554, 2006.

[Leordeanu and Hebert 05] M. Leordeanu and M. Hebert. "A spectral technique for correspondence problems using pairwise constraints." In *International Conference on Computer Vision*, pp. 1482–1489, 2005.

[Lessig and Fiume 08] C. Lessig and E. Fiume. "SOHO: Orthogonal and symmetric Haar wavelets on the sphere." *ACM Transactions on Graphics* 27:1 (2008), 4:1–4:11.

[Lévy 06] B. Lévy. "Laplace-Beltrami eigenfunctions towards an algorithm that "understands" geometry." In *Shape Modeling International*, pp. 13–20, 2006.

[Li et al. 12] S. Li, Q. Zhao, S. Wang, T. Hou, A. Hao, and H. Qin. "A Novel Material-aware Feature Descriptor for Volumetric Image Registration in Diffusion Tensor Space." In *European Conference on Computer Vision*, pp. 493–506, 2012.

[Lipman and Funkhouser 09] Y. Lipman and T. Funkhouser. "Möbius voting for surface correspondence." *ACM Transactions on Graphics* 28:3 (2009), 72:1–72:12.

[Lipman et al. 04] Y. Lipman, O. Sorkine, D. Cohen-Or, D. Levin, C. Rössl, and H.-P. Seidel. "Differential coordinates for interactive mesh editing." In *Shape Modeling International*, pp. 181–190, 2004.

[Lipman et al. 10] Y. Lipman, R. M. Rustamov, and T. A. Funkhouser. "Biharmonic Distance." *ACM Transactions on Graphics* 29:3 (2010), 27:1–27:11.

[Lippert and Gross 95] L. Lippert and M. H. Gross. "Fast Wavelet Based Volume Rendering by Accumulation of Transparent Texture Maps." *Computer Graphics Forum* 14:3 (1995), 431–444.

[Liu and Zhang 07] R. Liu and H. Zhang. "Mesh segmentation via spectral embedding and contour analysis." *Computer Graphics Forum* 26:3 (2007), 385–394.

[Liu et al. 06] Y. Liu, H. Zha, and H. Qin. "Shape Topics: A Compact Representation and New Algorithms for 3D Partial Shape Retrieval." In *IEEE*

Conference on Computer Vision and Pattern Recognition, pp. 2025–2032, 2006.

[Loop 87] C. Loop. "Smooth Subdivision Surfaces Based on Triangles." Master's thesis, University of Utah, 1987.

[Lounsbery et al. 97] M. Lounsbery, T. D. DeRose, and J. Warren. "Multiresolution analysis for surfaces of arbitrary topological type." *ACM Transactions on Graphics* 16:1 (1997), 34–73.

[Lowe 04] D. Lowe. "Distinctive image features from scale-invariant keypoints." *International Journal of Computer Vision* 60:2 (2004), 91–110.

[Lu et al. 09] J. Lu, J. Dorsey, and H. Rushmeier. "Dominant Texture and Diffusion Distance Manifolds." *Computer Graphics Forum* 28:2 (2009), 667–676.

[Luo et al. 09] C. Luo, I. Safa, and Y. Wang. "Approximating gradients for meshes and point clouds via diffusion metric." *Computer Graphics Forum* 28:5 (2009), 1497–1508.

[Maggioni and Mhaskar 08] M. Maggioni and H.N. Mhaskar. "Diffusion polynomial frames on metric measure spaces." *Applied and Computational Harmonic Analysis* 24:3 (2008), 329–353.

[Maggioni et al. 05] M. Maggioni, J. Bremer, R. R. Coifman, and A. Szlam. "Biorthogonal diffusion wavelets for multiscale representations on manifolds and graphs." In *SPIE Wavelets XI*, 5914, 5914, pp. 1–13, 2005.

[Maggioni 07] M. Maggioni. "Multiscale Analysis of Data Sets with Diffusion Wavelets." Citeseer, 2007.

[Mahadevan and Maggioni 05] S. Mahadevan and M. Maggioni. "Value function approximation with diffusion wavelets and laplacian eigenfunctions." In *Neural Information Processing Systems*, 2005.

[Malcolm et al. 07] J. Malcolm, Y. Rathi, and A. Tannenbaum. "A Graph Cut Approach to Image Segmentation in Tensor Space." In *IEEE Conference on Computer Vision and Pattern Recognition*, pp. 1–8, 2007.

[Mallat 89] S. G. Mallat. "A Theory for Multiresolution Signal Decomposition: The Wavelet Representation." *IEEE Transactions on Pattern Analysis and Machine Intelligence* 11:7 (1989), 674–693.

[Mallat 08] S. Mallat. *A Wavelet Tour of Signal Processing, Third Edition: The Sparse Way.* Academic Press, 2008.

[Mateus et al. 08] D. Mateus, R.P. Horaud, D. Knossow, F. Cuzzolin, and E. Boyer. "Articulated shape matching using laplacian eigenfunctions and unsupervised point registration." In *IEEE Conference on Computer Vision and Pattern Recognition*, pp. 1–8, 2008.

[Mémoli 09] F. Mémoli. "Spectral Gromov-Wasserstein distances for shape matching." In *Workshop on Non-Rigid Shape Analysis and Deformable Image Alignment*, pp. 256–263, 2009.

[Meyer et al. 02] M. Meyer, M. Desbrun, P. Schröder, and A. H. Barr. "Discrete differential-geometry operators for triangulated 2-manifolds." *Visualization and Mathematics* 3:7 (2002), 35–57.

[Meyer 92] Y. Meyer. *Wavelets and Operators.* Cambridge University Press, 1992.

[Meyer 01] Y. Meyer. *Oscillating Patterns in Image Processing and Nonlinear Evolution.* American Mathematical Society, 2001.

[Morlet et al. 82a] J. Morlet, G. Arens, E. Fourgeau, and D. Glard. "Wave Propagation and Sampling Theory - Part I: Complex Signal and Scattering in Multilayered Media." *GEOPHYSICS* 47:2 (1982), 203–221.

[Morlet et al. 82b] J. Morlet, G. Arens, E. Fourgeau, and D. Glard. "Wave Propagation and Sampling Theory - Part II: Sampling Theory and Complex Waves." *GEOPHYSICS* 47:2 (1982), 222–236.

[Narcowich et al. 96] F. J. Narcowich, P. Petrushev, and J. D. Ward. "Localized tight frames on spheres." *SIAM Journal of Mathematical Analysis* 38:2 (1996), 574–594.

[Nielson et al. 97] G. M. Nielson, I.-H. Jung, and J. Sung. "Haar wavelets over triangular domains with applications to multiresolution models for flow over a sphere." In *IEEE Visualization*, pp. 143–150, 1997.

[Novatnack and Nishino 07] J. Novatnack and K. Nishino. "Scale-dependent 3d geometric features." In *International Conference on Computer Vision*, pp. 1–8, 2007.

[Olsen et al. 07] L. Olsen, F. F. Samavati, and R. H. Bartels. "Multiresolution for curves and surfaces based on constraining wavelets." *Computers & Graphics* 31:3 (2007), 449–462.

[Osada et al. 01] R. Osada, T. Funkhouser, B. Chazelle, and D. Dobkin. "Matching 3d models with shape distributions." In *Shape Modeling International*, pp. 154–166, 2001.

[Osada et al. 02] R. Osada, T. Funkhouser, B. Chazelle, and D. Dobkin. "Shape distributions." *ACM Transactions on Graphics* 21:4 (2002), 807–832.

[Ovsjanikov et al. 08] M. Ovsjanikov, J. Sun, and L. Guibas. "Global intrinsic symmetries of shapes." *Computer Graphics Forum* 27:5 (2008), 1341–1348.

[Ovsjanikov et al. 09] M. Ovsjanikov, A. M. Bronstein, L. J. Guibas, and M. M. Bronstein. "Shape Google: a computer vision approach to invariant shape retrieval." In *ICCV Workshops on NORDIA*, pp. 320–327, 2009.

[Ovsjanikov et al. 10] M. Ovsjanikov, Q. Mérigot, F. Mémoli, and L. Guibas. "One point isometric matching with the heat kernel." *Computer Graphics Forum* 29:5 (2010), 1555–1564.

[Patanè and Falcidieno 10] G. Patanè and B. Falcidieno. "Multi-scale feature spaces for shape processing and analysis." In *Shape Modeling International*, pp. 113–123, 2010.

[Pauly and Gross 01] M. Pauly and M. Gross. "Spectral Processing of Point-Sampled Geometry." In *ACM SIGGRAPH*, pp. 379–386, 2001.

[Pauly et al. 03] M. Pauly, R. Keiser, and M. Gross. "Multi-scale feature extraction on point-sampled surfaces." *Computer Graphics Forum* 22:3 (2003), 281–289.

[Payan and Antonini 07] F. Payan and M. Antonini. "Temporal wavelet-based compression for 3D animated models." *Computers & Graphics* 31:1 (2007), 77–88.

[Raviv et al. 11] D. Raviv, A. M Bronstein, M. M Bronstein, R. Kimmel, and N. Sochen. "Affine-invariant diffusion geometry for the analysis of deformable 3D shapes." *Computers & Graphics* 35:3 (2011), 692–697.

[Reuter et al. 05] M. Reuter, F.-E. Wolter, and N. Peinecke. "Laplace-Spectra as Fingerprints for Shape Matching." In *ACM Symposium on Solid and Physical Modeling*, pp. 101–106, 2005.

[Reuter et al. 06] M. Reuter, F.-E. Wolter, and N. Peinecke. "Laplace-Beltrami Spectra as "Shape-DNA" of Surfaces and Solids." *Computer-Aided Design* 38:4 (2006), 342–366.

[Reuter et al. 09] M. Reuter, F.-E. Wolter, M. Shenton, and M. Niethammer. "Laplace-Beltrami eigenvalues and topological features of eigenfunctions for statistical shape analysis." *Computer-Aided Design* 41:10 (2009), 739–755.

[Reuter 10] M. Reuter. "Hierarchical shape segmentation and registration via topological features of Laplace-Beltrami eigenfunctions." *International Journal of Computer Vision* 89:2 (2010), 287–308.

[Roşca and Antoine 09] D. Roşca and J.-P. Antoine. "Locally Supported Orthogonal Wavelet Bases on the Sphere via Stereographic Projection." *Mathematical Problems in Engineering* 29.

[Rong et al. 08] G. Rong, Y. Cao, and X. Guo. "Spectral mesh deformation." *The Visual Computer* 24:7 (2008), 787–796.

[Ruggeri et al. 10] M. R. Ruggeri, G. Patanè, M. Spagnuolo, and D. Saupe. "Spectral-driven isometry-invariant matching of 3D shapes." *International Journal of Computer Vision* 89:2-3 (2010), 248–265.

[Rustamov 07] R. M. Rustamov. "Laplace-beltrami eigenfunctions for deformation invariant shape representation." In *Symposium on Geometry Processing*, pp. 225–233, 2007.

[Scheunders 02] P. Scheunders. "A Multivalued Image Wavelet Representation Based on Multiscale Fundamental Forms." *IEEE Transactions on Image Processing* 11:30 (2002), 568–575.

[Sfikas et al. 11] K. Sfikas, I. Pratikakis, and T. Theoharis. "ConTopo: Non-Rigid 3D Object Retrieval using Topological Information guided by Conformal Factors." In *Eurographics 2011 Workshop on 3D Object Retrieval*, pp. 25–32, 2011.

[Strang 86] G. Strang. *Introduction to Applied Mathematics.* Wellesley-Cambridge Press, 1986.

[Su et al. 09] Z. Su, H. Wang, and J. Cao. "Mesh Denoising Based on Differential Coordinates." In *Shape Modeling and Applications*, pp. 1–6, 2009.

[Sun et al. 09] J. Sun, M. Ovsjanikov, and L. Guibas. "A Concise and Provably Informative Multi-Scale Signature Based on Heat Diffusion." *Computer Graphics Forum* 28:5 (2009), 1383–1392.

[Sun et al. 10] J. Sun, X. Chen, and T. A. Funkhouser. "Fuzzy geodesics and consistent sparse correspondences for deformable shapes." *Computer Graphics Forum* 29:5 (2010), 1535–1544.

[Sunkel et al. 11] M. Sunkel, S. Jansen, M. Wand, E. Eisemann, and H.-P. Seidel. "Learning Line Features in 3D Geometry." *Computer Graphics Forum* 30:2 (2011), 267–276.

[Tabia et al. 10] H. Tabia, M. Daoudi, J. P. Vandeborre, and O. Colot. "Local visual patch for 3d shape retrieval." In *ACM Workshop on 3D Object Retrieval*, pp. 3–8, 2010.

[Tam and Lau 07] G. K. L. Tam and R. W. H. Lau. "Deformable Model Retrieval Based on Topological and Geometric Signatures." *IEEE Transactions on Visualization and Computer Graphics* 13:3 (2007), 470–482.

[Tangelder and Veltkamp 08] J. W. Tangelder and R. C. Veltkamp. "A survey of content based 3D shape retrieval methods." *Multimedia Tools and Applications* 39:3 (2008), 441–471.

[Tevs et al. 09] A. Tevs, M. Bokeloh, and M. Wand. "Isometric registration of ambiguous and partial data." In *IEEE Conference on Computer Vision and Pattern Recognition*, pp. 1185–1192, 2009.

[Tierny et al. 09] J. Tierny, J.-P. Vandeborre, and M. Daoudi. "Partial 3D Shape Retrieval by Reeb Pattern Unfolding." *Computer Graphics Forum* 28:1 (2009), 41–55.

[Toldo et al. 10] R. Toldo, U. Castellani, and A. Fusiello. "The bag of words approach for retrieval and categorization of 3D objects." *The Visual Computer* 26:10 (2010), 1257–1268.

[Torresani et al. 08] L. Torresani, V. Kolmogorov, and C. Rother. "Feature correspondence via graph matching." In *European Conference on Computer Vision*, pp. 596–609, 2008.

[Valette and Prost 04] S. Valette and R. Prost. "Wavelet-based multiresolution analysis of irregular surface meshes." *IEEE Transactions on Visualization and Computer Graphics* 10:2 (2004), 113–122.

[Vallet and Lévy 08] B. Vallet and B. Lévy. "Spectral geometry processing with manifold harmonics." *Computer Graphics Forum* 27:2 (2008), 251–260.

[Vaxman et al. 10] A. Vaxman, M. Ben-Chen, and C. Gotsman. "A multiresolution approach to heat kernels on discrete surfaces." *ACM Transactions on Graphics* 29:4 (2010), 121:1–121:10.

[Vlasic et al. 08] D. Vlasic, I. Baran, W. Matusik, and J. Popović. "Articulated mesh animation from multi-view silhouettes." *ACM Transactions on Graphics* 27:3 (2008), 97:1–97:9.

[Wang and Mahadevan 09] C. Wang and S. Mahadevan. "Multiscale Analysis of Document Corpora based upon Diffusion Models." In *International Joint Conference on Artificial Intelligence*, pp. 1592–1597, 2009.

[Wang and Vemuri 04] Z. Wang and B. C. Vemuri. "An Affine Invariant Tensor Dissimilarity Measure and Its Applications to Tensor-Valued Image Segmentation." In *IEEE Conference on Computer Vision and Pattern Recognition*, pp. 228–233, 2004.

[Wang et al. 07] H. Wang, K. Qin, and H. Sun. "$\sqrt{3}$-Subdivision-Based Biorthogonal Wavelets." *IEEE Transactions on Visualization and Computer Graphics* 13:5 (2007), 914–925.

[Wang et al. 08] S. Wang, X. Gu, and H. Qin. "Automatic non-rigid registration of 3d dynamic data for facial expression synthesis and transfer." In *IEEE Conference on Computer Vision and Pattern Recognition*, 2008.

[Wang et al. 11a] S. Wang, T. Hou, Z. Su, and H. Qin. "Multi-scale anisotropic heat diffusion based on normal-driven shape representation." *The Visual Computer* 27:6-8 (2011), 429–439.

[Wang et al. 11b] S. Wang, T. Hou, Z. Su, and H. Qin. "Diffusion Tensor Weighted Harmonic Fields for Feature Classification." In *Pacific Conference on Computer Graphics and Applications*, pp. 93–98, 2011.

[Wu et al. 08] C. Wu, B. Clipp, X. Li, J.-M. Frahm, and M. Pollefeys. "3d model matching with viewpoint-invariant patches (vip)." In *IEEE Conference on Computer Vision and Pattern Recognition*, 2008.

[Wu 04] F. Y. Wu. "Theory of resistor networks: The two-point resistance." *Journal of Physics A* 37 (2004), 6653–6673.

[Young 87] R. A. Young. "The Gaussian derivative model for spatial vision: I. Retinal mechanisms." *Spatial Vision* 2:4 (1987), 273–293.

[Yu et al. 03] M. Yu, I. Atmosukarto, W. K. Leow, Z. Huang, and R. Xu. "3D Model Retrieval With Morphing-Based Geometric and Topological Feature Maps." In *IEEE Conference on Computer Vision and Pattern Recognition*, pp. 656–661, 2003.

[Yu et al. 04] Y. Yu, K. Zhou, D. Xu, X. Shi, H. Bao, B. Guo, and H.-Y. Shum. "Mesh editing with Poisson-based gradient field manipulation." In *ACM SIGGRAPH*, pp. 644–651, 2004.

[Zaharescu et al. 09] A. Zaharescu, E. Boyer, K. Varanasi, and R. Horaud. "Surface feature detection and description with applications to mesh matching." In *IEEE Conference on Computer Vision and Pattern Recognition*, pp. 373–380, 2009.

[Zeng et al. 10] Y. Zeng, C. Wang, Y. Wang, X. Gu, D. Samaras, and N. Paragios. "Dense non-rigid surface registration using high-order graph matching." In *IEEE Conference on Computer Vision and Pattern Recognition*, pp. 382–389, 2010.

[Zhang et al. 04] L. Zhang, Noah Snavely, B. Curless, and S. M. Seitz. "Spacetime Faces: High-resolution capture for modeling and animation." *ACM Transactions on Graphics* 23:3 (2004), 548–558.

[Zhang et al. 10] J. Zhang, J. Zheng, and J. Cai. "A diffusion approach to seeded image segmentation." In *IEEE Conference on Computer Vision and Pattern Recognition*, pp. 2125–2132, 2010.

[Zhong et al. 12] M. Zhong, T. Hou, and H. Qin. "A Hierarchical Approach to High-Quality Partial Shape Registration." In *International Conference on Pattern Recognition*, pp. 113–116, 2012.

[Zou et al. 08] G. Zou, J. Hua, M. Dong, and H. Qin. "Surface matching with salient keypoints in geodesic scale space." *Computer Animation and Virtual Worlds* 19:3-4 (2008), 399–410.

[Zou et al. 09] G. Zou, J. Hua, Z. Lai, X. Gu, and M. Dong. "Intrinsic geometric scale space by shape diffusion." *IEEE Transactions on Visualization and Computer Graphics* 15:6 (2009), 1193–1200.

Index